The Regulation of
Dietary Supplements

The Regulation of Dietary Supplements
A Historical Analysis

Stephen J. Pintauro

CRC Press
Taylor & Francis Group
Boca Raton London New York

CRC Press is an imprint of the
Taylor & Francis Group, an **informa** business

CRC Press
Taylor & Francis Group
6000 Broken Sound Parkway NW, Suite 300
Boca Raton, FL 33487-2742

First issued in paperback 2019

ISBN-13: 978-0-376-90194-3 (pbk)
ISBN-13: 978-1-138-33754-1 (hbk)

Visit the Taylor & Francis Web site at
http://www.taylorandfrancis.com

and the CRC Press Web site at
http://www.crcpress.com

Dedicated to
Sallie
Nicholas
Emily

Contents

Preface

When it comes to the topic of food and nutrition, many of us have strong opinions. How much should we eat? What is healthy to eat? What is unhealthy to eat? The list can go on and on. But perhaps one of the most hotly debated and controversial nutrition topics among both health professionals and the general public is the issue of dietary supplements. As a nutritional scientist and university professor, I often found myself addressing various aspects of the topic in many of the classes I have taught over the past 37 years. But regardless of the context in which I was covering the topic, inevitably, some aspect of dietary supplement *regulation* would find its way into the discussion. So early in my academic career, I began teaching a course on food regulation, in which the topic of dietary supplements was a major component. While developing this course in the late 1980s, I attended a workshop on the topic of "food health claims" conducted by the Food and Drug Law Institute in Washington, DC. The audience and speakers at the workshop were an interesting mix of scientists, industry representatives, and Food and Drug Administration (FDA) regulators. Discussion of the topic of health claims on foods and dietary supplements among these various stakeholder workshop participants and speakers became very heated at times. It quickly became apparent to me that there was a lot of fascinating history to this debate, and telling the story of this history became an integral part of my Food Regulation class. Of course, the story did not end with the 1980s. Dietary supplement regulation continued to evolve in many significant ways over the next 30 years. My goal in writing this book is to tell the complete story of dietary supplement regulation, from the federal government's early involvement at the beginning of the 20th century until today. In telling the story, I wanted to highlight the challenges faced by lawmakers, federal regulators (particularly the FDA), and the courts when trying to balance the interests and rights of all stakeholders. I also wanted to demonstrate the role that science plays, sometime successfully and sometimes unsuccessfully, in directing public policy. While there are many examples of scientific issues included in the book, one does not need a strong background in science to hopefully find the book interesting and useful. Neither does one need to be a lawyer to understand the legal aspects of this story. In the early chapters of the book, I include enough background on the relevant science and government function to allow the book to be understandable to anyone with a general interest in nutrition and health, food law, the FDA, or science and public policy. This would include college students (perhaps even high school students), professors/teachers, food industry representatives, consumer advocates, or any others with an interest in the history of dietary supplement regulation.

About the author

Dr. Stephen J. Pintauro earned his MS in nutritional biochemistry and his PhD in food science from the University of Rhode Island, Kingston, Rhode Island. He is currently an associate professor in the department of nutrition and food sciences at the University of Vermont, Burlington, Vermont, where he teaches courses in advanced nutrition, research methods in nutrition and food sciences, and food regulation. Dr. Pintauro is the recipient of two prestigious teaching awards while at the University of Vermont: the College of Agriculture and Life Sciences Joseph E. Carrigan Award in 2000 and the University's top teaching award, the Kroepsch-Maurice Award in 2015. He is a member of the American Society for Nutrition, the Academy of Nutrition and Dietetics, the eLearning Guild, and the Institute of Food Technologists (IFT), including the IFT Food Law and Regulation Division. For the past 20 years, Dr. Pintauro's research and scholarship efforts have focused on the development and testing of nutrition and food science-related information technology applications. His research has been funded through grants from the United States Department of Agriculture, the National Science Foundation, and the American Cancer Society. He and his wife, Sallie, split their time between their homes in Charlotte, Vermont, and Bonita Springs, Florida.

chapter one

Introduction

Any regular viewer of evening news programs on television (TV) is certainly aware of the pharmaceutical theme that pervades the commercial breaks of these shows. Clearly, the manufacturers of many health-related products and services consider viewers of these programs to be prime targets for their advertising dollars. On just a recent 30-minute evening news broadcast, I noted commercials for the following:

Phillips Colon Health (probiotic capsules)
Prevnar13 (pneumococcal pneumonia vaccine)
Humira (rheumatoid arthritis)
One-A-Day Men's (multivitamin)
Zantac (for heartburn)
Aleve (pain reliever)
Prevagen (for brain health and memory)
Linzess (for chronic idiopathic constipation)

It seems that our daily dose of evening news includes a virtual tour of your local pharmacy. I suppose this shouldn't be surprising. After all, older Americans represent the largest proportion of network news viewers. A fairly recent Pew Research Center survey[1] reported that approximately 40% of individuals aged 65 and above are regular viewers of evening news, compared with only 11% of those between the ages of 18 and 29 and 26% of those between the ages of 30 and 49. It does not take a Madison Avenue genius to recognize that these older viewers are also the folks most likely to purchase these products. But a closer look at this list of TV ads reveals some additional interesting information. Of the eight ads, three are for prescription drugs (Prevnar13, Humira, Linzess), two are for over-the-counter (OTC) drugs (Zantac and Aleve), and three are for dietary supplements (Philips Colon Health, Prevagen, and One-a-Day Men's multivitamin). So, these ads represent three different categories of products: prescription drugs, OTC drugs, and dietary supplements.

The federal laws and regulations that oversee the manufacture, labeling, and advertising of these three categories of products differ in many important ways. One of the most significant differences among these three

---

[1] 2012 News Consumption Survey, May 9–June 3, 2012, Pew Research Center.

categories of substances is that, for both prescription and OTC drugs, the laws and regulations require that the manufacturers provide the U.S. Food and Drug Administration (FDA) with scientific data, establishing both the safety and effectiveness of these products before they are permitted for sale. As we shall see in subsequent chapters of this book, no such premarket safety and effectiveness testing is required for dietary supplements. Yet, while the principal focus of this book is on the regulation of dietary supplements, we will necessarily find ourselves also dealing with issues related to drug regulation. This is because the distinction between what is a "drug" and what is a "dietary supplement" is not always easy for the consuming public to discern. Let us start by simply taking a closer look at the narratives in the three dietary supplement TV ads mentioned earlier. The first is the commercial for Phillips Colon Health probiotic supplement.

[Female spokesperson wearing a Philips T-shirt that reads "Regular and proud of it" climbs onto the upper deck of a bus, holding a package of Philips Colon Health Probiotic Capsule supplement, and asks the passengers on the bus:]

"Anyone have occasional constipation, diarrhea, gas, bloating?"

One husband points shyly at his wife.

Another passenger says, "Yes."

[Female Philips spokeswoman]: "One Philips Colon Health Probiotic capsule each day helps defend against these digestive issues—with three strains of good bacteria."

[Female Philips spokeswoman then begins to pass out boxes of the Philips Probiotic supplement to the passengers.]

[Spokeswoman voiceover]: "Live the regular life. Philips."

The second commercial is for One-A-Day Men's 50+ multivitamin.

[Male voiceover. Video of men and boys camping and cooking outdoors.] "Americans. 83% try to eat healthy. Yet up to 90% fall short in getting key nutrients from food alone."

[Video of man holding bottle of vitamins, Male voiceover.] "Let's do more. Add One-A-Day Men's. Complete with key nutrients we may need."

[Man looking at bottle of vitamins. Male voiceover.] "Plus heart health support with B vitamins. One-A-Day Men's. In gummies and tablets."

And the third commercial is for Prevagen brain and memory supplement.

[Male voiceover, with animations of brain and nerve activity]: "Your brain is an amazing thing. But as you get older it naturally begins to change, causing a lack of sharpness, or even trouble with recall. Thankfully, the breakthrough in Prevagen helps your brain, and actually improves memory. The secret is an ingredient originally discovered in jellyfish. In clinical trials, Prevagen has been shown to improve short-term memory. Prevagen. The name to remember."

As you read these transcripts of the TV ads, you may have noticed that there is something very different between these ads and those for the prescription or OTC drugs. Unlike the drug commercial advertisements, there is no mention of a specific disease or medical condition in the advertisements of these dietary supplements. This is not unintentional. According to federal law, any product label that makes a claim related to its *"intended—use in the diagnosis, cure, mitigation, treatment, or prevention of disease in man or other animals"* is, by legal definition, considered to be a drug (with a few notable exceptions that we will discuss later in this book). And no drug, whether it be prescription or OTC, can be sold in the United States until the manufacturer provides sufficient scientific evidence to the FDA that it is both safe *and* effective. Establishing safety and effectiveness of any drug can typically cost the manufacturer many millions of dollars in animal testing and human clinical testing and often takes many years to complete. Can the manufacturer of Phillips Colon Health capsule afford to do the necessary testing to prove that the product is safe? Well, the parent company of this product is Bayer AG, a very large multinational pharmaceutical company. So they certainly have the resources to do the necessary safety testing. But that would likely greatly increase the cost of the product to the consumer. Besides, there is the additional requirement of establishing the effectiveness of the product. Does Phillips Colon Health really improve issues related to constipation, bloating, and gas? Again, if this product met the legal definition of a drug, Bayer would need to provide the FDA with reliable and convincing scientific data to substantiate the product's labeling and advertising claims for these gastrointestinal issues. Even if a company such as Bayer could afford this type of scientific testing, for many—if not most—dietary supplements, proving effectiveness may be a difficult task. Some of the reasons for this will be addressed later in this book.

Yet, dietary supplement manufacturers need to communicate what they believe are the health—and even disease—preventing or treating benefits of their products to the consumer. How can they do this and still avoid their products being legally classified as drugs? Let's consider this by taking another look at the Prevagen ad transcript. Notice the line, *"Researchers have discovered a protein that actually supports healthier brain function."* And the line, *"For support of healthier brain function, a sharper mind, and clearer thinking."*

And notice similar claims in the Phillips Colon Health and One-A-Day Men's Multivitamin ads. For Phillips Colon Health the ad states, *"Help defend against those digestive issues."* In the One-A-Day ad, notice the line, *"Plus heart health support with B vitamins."* These types of statements are permitted because they do not make a specific disease claim. Rather, they make claims related to the *structure or function* of the body. The manufacturers intend and hope that consumers will make the link between these "structure and function" claims and the possible benefits to specific diseases. So, the One-A-Day ad states that the product offers *"...heart health support with B vitamins."* But it does not state that it helps in the treatment or prevention of heart disease. And the Prevagen ad states, *"For support of healthier brain function..."* But it does not state that it helps prevent dementia or related brain diseases such as Alzheimer's disease.

And there is one other important legal requirement in the advertising and labeling of these dietary supplements. If you watch the actual TV ads closely, you may notice a small font disclaimer that appears briefly at the bottom of the screen. This disclaimer states

These statements have not been evaluated by the FDA.
This product is not intended to treat, cure, or prevent any disease.

But why aren't dietary supplements held to the same standards of safety and effectiveness that the FDA requires of all drugs? This is because in 1994 the U.S. Congress passed and President Clinton signed into law the Dietary Supplement Health and Education Act. Throughout this book, we will simply refer to this law as the DSHEA. This law radically altered the way dietary supplements are regulated in the United States. Among the law's provisions is one that permits the manufacturers of dietary supplements to make "structure or function" claims, such as the ones for the products described earlier.

There are many other important provisions to this law, and we will explore these in detail in this book. However, this book is not intended to simply review the major provisions of the DSHEA. Rather, it is intended to explore the roles that many different factors and issues played in the nearly 100-year history of dietary supplement regulation, culminating in the passage of the DSHEA in 1994, and some important changes that have occurred since. Some of the issues to be explored include

How big is the dietary supplement industry in the United States? We all know that many Americans take dietary supplements. But how is that reflected in the growth and size of the industry? How much are Americans spending on dietary supplements, and how much money are manufacturers making in the sale of these products?

Why do people take dietary supplements? This may seem like a simple question, but research has shown that the reasons will vary based on many factors, such as gender, age, health status, education, socioeconomic status, and more. An understanding of the role that these factors play in driving consumer demand for dietary supplements is important to understanding the interaction between science and the enactment of public policy.

Are dietary supplements effective? Of course, there is no simple yes or no answer to this question. Although the DSHEA does not require that manufacturers of dietary supplements prove their "structure or function" claims, there is nevertheless considerable interest among consumers regarding the effectiveness of the products they are purchasing. However, there are literally thousands of different dietary supplement products on the market, making it virtually impossible to scientifically evaluate the effectiveness of each. And how do we define "effective"? If 5 out of 100 users of a probiotic dietary supplement experience a slight improvement in, for example, "gastrointestinal discomfort," is that a sufficient indication of effectiveness? There is no absolute threshold at which a substance can officially be considered "effective." But we can apply some basic scientific principles to help us make determinations about the relative effectiveness and the strength of the scientific evidence related to effectiveness. Consideration of these scientific principles and the role that they should play in the regulation of dietary supplements have driven much of the debate between federal regulators (such as the FDA), legislators, manufacturers, and the consuming public.

Are dietary supplements safe? As with the "effectiveness" issue, the DSHEA does not require premarket safety testing of dietary supplements. Yet, the safety of dietary supplements has been a major consideration in the long history of dietary supplement laws and regulations. Evaluating the safety of dietary supplements presents some unique scientific and regulatory challenges. All dietary supplements are different and present different levels of risk with regard to toxicity. As with any substance, whether it be a dietary supplement, a drug, or any food item, its toxicity is fundamentally related to the amount consumed, or more precisely, the dose. If you consume high-enough levels of any substance, it will be toxic. Hence, there is no such thing as absolute safety (or zero risk) for any substance. So rather than posing the question, "Are dietary supplements safe?" it is more appropriate to ask, "What is the risk from consuming a particular dietary supplement?" "Safety" is not something that can be scientifically measured. However, risk is something that can be measured or estimated. Research and consumer experiences have clearly shown that some dietary supplements present unacceptable risks of harm. We will examine some important examples of these in later chapters, which bring us to the next issue.

Are dietary supplements adequately regulated? This question certainly relates to the earlier questions, as consumers of dietary supplements have a right to the very reasonable expectation that the supplements that they are consuming are "safe." Does the DSHEA meet this expectation? And what about other aspects of dietary supplement regulation? For example, should manufacturers of dietary supplements be required to substantiate their labeling and advertising claims? Should there be limits on the types and combinations of substances that can be legally classified as dietary supplements?

What are some of the historical, scientific, political, and societal factors that led to the passage of the DSHEA? The passage of this law in 1994 did not occur spontaneously or without a historical context. For nearly 100 years, the FDA (or its predecessor federal enforcement agency) has been aggressively pursuing an enforcement strategy designed to reign in what it considered to be serious problems in the dietary supplement industry. These problems covered the gamut from safety issues to consumer fraud issues. A substantial portion of this book is devoted to a systematic analysis of this historical context and the role it played in the passage of the DSHEA.

My intent in writing this book is to tell the story of dietary supplement regulation in a way that is understandable and useful to anyone with an interest in the topic, regardless of whether or not they have a background in food science, nutrition, law, or public policy. With that goal in mind, I think it is important to lay out some background to the topic, beginning with an overview of the extent of the use of dietary supplements in the United States and the reasons for their popularity. In addition, how much do consumers truly understand regarding the federal government's role in the regulation of the safety and effectiveness of dietary supplements?

chapter two

Dietary supplement use in the United States

It is estimated that Americans spend more than $30 billion on dietary supplements each year.[1] To put that figure in some perspective, consider how it compares with yearly consumer purchases of other "food-related" products. Americans spend approximately $11 billion per year on coffee, $65 billion per year on soft drinks, $96 billion per year on beer, $117 billion per year on fast food, and $478 billion per year on groceries.[2] So dietary supplements are right up there with the biggest categories of food-related purchases, representing as much as 5% of an individual's average food spending. Of course, since many individuals do not purchase any dietary supplements, while others regularly purchase many different dietary supplements, the range in the amount of money that people spend on dietary supplements will vary enormously. So what are the actual statistics with regard to dietary supplement use? Who is using dietary supplements and why are they using them? Interestingly, there have actually been many scientific studies conducted over the past several decades designed to answer these questions.

Let's start with a quick overview of how dietary supplement use has changed over time in the United States. The graph below illustrates the percentage of the surveyed population that has consumed a dietary supplement within the past month at the time the surveys were administered.

The data used to construct the above chart were obtained from published studies of dietary supplement use trends[3,4] and based on the National Health and Nutrition Examination Surveys (NHANES) conducted since 1974. NHANES is an ongoing (since 1971) research survey program designed to assess the health and nutrition status of adults and children in the United States. It is administered by the National Center for Health Statistics, which is part of the Centers for Disease Control and

[1] Garcia-Cazarin, ML, EA Wambogo, KS Regan, CD Davis. Dietary Supplement Research Portfolio at the NIH, 2009–2011. *J. Nutr.*, 144: 414–418 (2014).
[2] By the Numbers: How Americans Spend Their Money. (http://mentalfloss.com/article/31222/numbers-how-americans-spend-their-money).
[3] Briefel, RR and CL Johnson. Secular Trends in Dietary Intake in the United States. *Ann. Rev. Nutr.*, 24: 401–431 (2004).
[4] Kantor, ED, CD Relm, M Du, E White, EL Giovannucci. Trends in Dietary Supplement Use among US Adults from 1999–2012. *JAMA*, 316(14): 1464–1474 (2016).

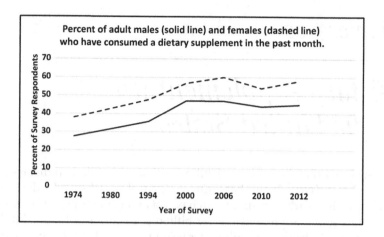

Prevention. Most of the information presented in this chapter regarding the use of dietary supplements by Americans was obtained from published studies that drew their results from data obtained from these NHANES surveys.

There are some interesting things to note about the above chart. First, it is important to understand the implications of the way NHANES obtained the dietary supplement use data. As the chart indicates, they asked respondents whether they used a dietary supplement in the past month. This would be likely to capture good data on regular users of dietary supplements, but may miss important data on occasional users of dietary supplements who simply did not use any in the month before the survey.

You will also notice that there is a consistent gender difference in dietary supplement use over the nearly 40-year period represented in the graph. Approximately 10% more adult women take dietary supplements, compared with adult men, and this difference has not changed much over the past 40 years. This gender difference is likely due to a number of factors. For example, women (particularly postmenopausal women) are often encouraged through mass media or by their health professionals to supplement their diets with calcium and vitamin D to help reduce their risk of osteoporosis. And women of child-bearing age are encouraged to take folic acid supplements to reduce the risk of neural tube birth defects if they become pregnant as well as iron supplements to reduce the risk of iron-deficiency anemias.

Another interesting point to note about this chart is the gradual but steady rise in supplement use that occurred from 1974 to 1994, followed by a steeper rise from 1994 to 2000, and then essentially no further rise in supplement use through 2012. One explanation for the early rise may be related to the way NHANES asked the questions. Through the mid-1980s, NHANES questions regarding dietary supplement use were limited to vitamin and/or mineral supplements. After 1988, dietary supplements use was expanded to include other products such as fish oil supplements, glucosamine, herbs, and botanicals. In addition, the rise in dietary supplement use observed around 1994 may be related to the passage in that year of the Dietary Supplement Health and Education Act (DSHEA), which granted the dietary supplement industry much more freedom with regard to the types of claims they could make on labels and in advertising. The leveling off in dietary supplement use observed from around the year 2000 to the present has been confirmed in numerous studies, with approximately 50%–55% of American adults (male and female combined) reporting that they have used a dietary supplement in the past month. Of course, this is an average figure and will vary based on other demographic characteristics.

Let's now take a look at some of the demographic factors associated with dietary supplement use. All of the following results are from the recent NHANES analysis study of Kantor et al.[5] The basic demographic patterns observed in the recent NHANES analysis are remarkably consistent with many previous surveys of dietary supplement use. The first chart below illustrates the relationship between dietary supplement use and age.

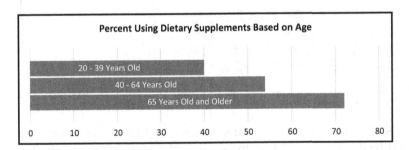

Perhaps not surprisingly, as people age and are confronted with the many health issues associated with aging, they are more likely to explore the use of dietary supplements to potentially help with these health issues.

In the following chart, you can see the relationship between level of education and dietary supplement use.

[5] Ibid.

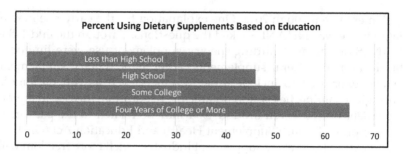

Clearly, the more education individuals have, the more likely they are to use dietary supplements. For those with less than a high school education, dietary supplement use is only 37%, while for those with a college education or more, dietary supplement use increases to 65%. Some may consider this to be a bit counterintuitive, as we may associate higher levels of education with the ability to better understand, interpret, and apply the scientific consensus information regarding the general lack of effectiveness for many dietary supplements. We will explore this issue in detail later in this chapter.

Finally, the chart below illustrates the relationship between self-reported health status and dietary supplement use.

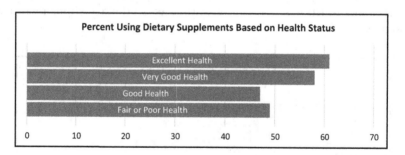

So, individuals who consider themselves to be in excellent or very good health are more likely to be users of dietary supplements. This is a very interesting observation, as other researchers have confirmed that these individuals are, in fact, more likely to consume a healthy diet, maintain a healthy weight, exercise regularly, and not smoke.[6,7] Dietary supplement users are indeed more healthy, or at least subscribe to a healthier

---

[6] Dickenson, A and D MacKay. Health Habits and Other Characteristics of Dietary Supplement Users: A Review. *Nutr. J.*, 13: 14 (2014).
[7] Kofoed, CLF, J Christensen, LO Dragsted, A Tjonneland, N Roswall. Determinants of Dietary Supplement Use – Healthy Individuals Use of Dietary Supplements. *Brit. J. Nutr.*, 113: 1993–2000 (2015).

lifestyle. This also makes it more difficult to evaluate the role that dietary supplements may play in their health status, as these other healthy diet and physical activity behaviors may mask any possible benefits from the supplements.

Dr. Regan Bailey from the Office of Dietary Supplements at the National Institutes of Health (NIH) and her colleagues used an analysis of the NHANES data to identify the reasons people take dietary supplements.[8] The nine most commonly reported reasons are indicated in the chart below.

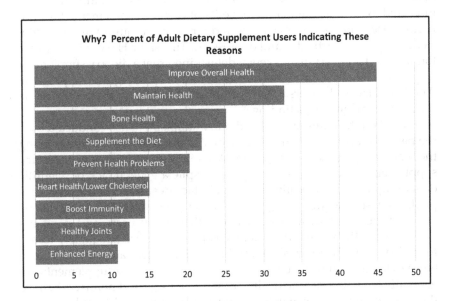

In this study, people were free to list as many reasons as they liked. But it is interesting to note that the most commonly reported reasons were related to general health (improve overall health, maintain health, supplement the diet, prevent health problems, boost immunity, enhance energy), as opposed to specific health conditions (bone health, heart health, healthy joints). However, these reasons varied somewhat based on age and gender. For example, older adults more often reported reasons related to specific health conditions such as heart health or bone health. And women tended to report use of dietary supplements more for bone health, colon health, and enhance energy, while men were reported to use more for heart health and mental health.

[8] Bailey, RL, JJ Gahche, PE Miller, PR Thomas, JT Dwyer. Why US Adults Use Dietary Supplements. *JAMA Intern. Med.*, 173(5): 355–367 (2013).

The belief in both the effectiveness and safety of dietary supplements is very strong among users and, even surprisingly, strong among nonusers. In a study published by Harvard University scientists in 2001,[9] investigators found that 85% of regular users of dietary supplements believed that they were good for health and well-being, while 34% of nonusers felt this way. While this may not seem surprising, consider that this study also found that 77% of regular users believed that dietary supplements were helpful for the common cold and 48% of nonusers agreed. This belief persists today despite over 40 years of research that has failed to show any beneficial role of dietary supplements in preventing or treating the common cold. Similar beliefs in the benefits of dietary supplements for both regular users and nonusers were found for cancer, acquired immunodeficiency syndrome, arthritis, and depression. This same Harvard study also found that if a government agency determined that a dietary supplement was ineffective, 72% of regular users would keep using the supplement. Another Harvard University study published in 2013 by some of the same scientists confirmed this finding.[10]

Based on the many studies of dietary supplement use conducted over the past 30 years, we now have a reasonably good picture of who is using them and why they are using them. What is still less clear is why dietary supplement use remains so prevalent, despite a general lack of scientific data to support their effectiveness, and in fact, despite several recent well-publicized studies that suggest that some supplements increase users' risk of harm or disease. We do know that many users and nonusers alike incorrectly assume that dietary supplements are adequately tested for safety and effectiveness before being sold. Users and nonusers also believe that the claims made in the advertising of dietary supplements are generally true.[8] But paradoxically, these same users and nonusers believe that new dietary supplements should not be sold until they have been "tested" by the U.S. Food and Drug Administration (FDA), and that more should be done by the government to ensure that dietary supplements are not harmful and that their advertising claims are true.[8] These findings suggest that the general public (both users and nonusers of dietary supplements) harbor some important misconceptions regarding both the safety and effectiveness of dietary supplements, and the role that the government is allowed to play in regulating this industry and its products.

The extent of the misunderstanding of dietary supplement regulation by the general public has been investigated in recent years. In 2011,

---

[9] Blendon, RJ, CM DesRoches, JM Benson, M Brodie, DE Altman. Americans' Views on the Use and Regulation of Dietary Supplements. *Arch. Intern. Med.*, 161: 805–810 (2001).

[10] Blendon, RJ, JM Benson, MD Botta, KJ Weldon. Users' Views of Dietary Supplements. *JAMA Intern. Med.*, 173(1): 74–76 (2013).

Dr. Tonya Dodge, currently an associate professor of psychology at George Washington University, surveyed 185 college students regarding their understanding of the topic.[11] The table below presents some of the results from this study.

Percentage of those surveyed answering correctly	% Correct
The FDA must approve a dietary supplement before it can be sold in nutrition stores	35.7
Manufacturers must provide scientific evidence of product safety to FDA before it can be sold	31.4
The FDA is responsible for ensuring that supplements are safe before they can be sold to consumers	25.0
The FDA is required to approve or disapprove dietary supplements	40.0
Manufacturers must present scientific evidence of product effectiveness to the FDA before the supplement can be sold	57.8
Dietary supplements sold in nutrition stores have been approved by the FDA	46.5
The FDA is responsible for analyzing the content of all dietary supplements that are sold to consumers	46.5

It is clear that at least among this particular population of consumers (college students), there is a significant lack of knowledge regarding dietary supplement regulation. However, this lack of knowledge extends well beyond college students. A study of physicians' understanding of dietary supplement regulation found somewhat similar results, with 37% of those physicians surveyed being unaware that dietary supplements did not require FDA approval before being sold.[12] Approximately the same percentage of physicians was unaware that efficacy and safety data are not required before dietary supplements are put on the market.

With so many people using dietary supplements for so many reasons, notwithstanding the many misconceptions regarding effectiveness, safety, and federal regulation, it is perhaps not surprising that the industry growth has exploded in recent decades. But how big is the dietary supplement industry? An exact figure is difficult to come by. A recent report

[11] Dodge, T, D Litt, A Kaufman. Influence of the Dietary Supplement Health and Education Act on Consumer Beliefs about the Safety and Effectiveness of Dietary Supplements. *J. Health Comm.*, 16: 230–244 (2011).
[12] Ashar, BH, TN Rice, SD Sisson. Physicians' Understanding of the Regulation of Dietary Supplements. *Arch. Intern. Med.*, 167: 966–969 (2007).

from the Office of Dietary Supplements at the NIH estimated total sales of dietary supplements in 2014 at $36.7 billion. However, the NIH obtained that figure from a dietary supplement trade organization publication (*The Nutrition Business Journal*), part of whose mission is to promote the dietary supplement industry. So this $36.7 billion sales estimate for 2014 probably represents the high side of the sales estimate. Yet even if the real sales figure was much smaller than this, it would still reflect a very large and profitable industry. And this is just U.S. sales. The global market for dietary supplements is expected to reach approximately $275 billion in the next 7 years.[13]

Dietary supplement sales are only part of the picture when considering the impact of this industry on U.S. and global economies. The Council for Responsible Nutrition, another pro-supplement trade organization, estimates that the dietary supplement industry contributes over $120 billion to the U.S. economy.[14] This estimate is based in part on the economic impact of the industry, creating over 750,000 jobs nationwide, paying $38.4 billion in wages, and contributing nearly $15 billion in federal and state business taxes (not including sales taxes on products).

The enormous profits to be made in the dietary supplement marketplace have not gone unnoticed by the big pharmaceutical drug companies. While "Big Pharma" has always had a share of the dietary supplement market, it has pursued this space much more aggressively in recent years. For example, One-A-Day vitamins is owned by the multinational chemical and pharmaceutical company, Bayer. Proctor & Gamble recently purchased the supplement company, New Chapter. Pfizer Pharmaceuticals acquired the Centrum line of supplement products as part of its takeover or Wyeth in 2009. And in 2012, Pfizer also acquired Alacer Corp., the maker of Emergen-C powdered drink dietary supplements. So, while the supplement industry as a whole likes to market its products as the "natural alternative" to dangerous and expensive drugs, it is often the drug companies themselves that are selling these products.

There is another interesting irony in this curious relationship between Big Pharma and the dietary supplement industry. Dietary supplement industry lobbying and advocacy groups are frequently suggesting that the FDA is mounting an all-out assault on your right to free access to the supplements of your choice. One need only conduct a quick Google search to find hundreds of links to sites that warn of impending FDA "crack downs" that would restrict sales to only those products that are "preapproved" by the FDA. While there is no factual basis for these

---

[13] Grand View Research. June 2016. (http://www.grandviewresearch.com/press-release/global-dietary-supplements-market).
[14] Council for Responsible Nutrition. Press Release. June 2016. (http://crn-archived-site.com/CRNPR16-HealthyProductsSupportHealthyEconomy060816.html).

concerns, these websites will often suggest that the FDA is being pressured to pursue more stringent regulation of the sale of dietary supplements by Big Pharma. Presumably, Big Pharma is out to protect its very profitable prescription and over-the-counter drug lines from competition from the much less expensive and "safer" dietary supplement alternatives. However, what these website authors fail to appreciate is the fact that the large drug companies also stand to lose if the FDA was ever able to more strictly regulate the supplement industry. In fact, it is precisely the relative unregulated nature of the supplement industry that likely attracted Big Pharma's expansion into this market.

Still, it is certainly true and no secret that the FDA, in its capacity as an Executive Branch enforcement agency, would like to more strictly regulate dietary supplements. It is also true that most reputable non-industry-affiliated public health advocacy groups, such as the American Medical Association, the American College of Physicians, the American College of Medical Toxicology, the American Academy of Clinical Toxicology, and the Center for Science in the Public Interest, and others, support the FDA's efforts in this regard. But the influence and impact that these groups can have on the ability of the FDA (or other agencies of the Executive Branch) to more strictly regulate this industry pales in comparison to the influence that industry-lobbying groups have on the legislative branch of our government. The simple fact is that the dietary supplement industry has very powerful friends in Congress. And these friends in Congress have been very effective at thwarting the efforts of the FDA to more strictly regulate this industry.

One of the industry's staunchest allies in Congress is Republican Senator Orrin Hatch of Utah. He was the chief architect and proponent of the successful passage of the DSHEA in 1994, which we will discuss in detail later in this book. According to a 2011 *New York Times* article,[15] Utah manufactures approximately 25% of the vitamin and other supplements produced in the United States, and it is the state's third largest industry. It is therefore not surprising that Senator Hatch is looking out for the industry's best interests. In return, the Utah senator has received hundreds of thousands of dollars in campaign contributions from industry political action committees (PACs) and executives. Interestingly, Senator Hatch recently announced that he will not seek reelection in November of 2018. How much this will impact legislative efforts related to the dietary supplement industry is unclear. But the congressional lobbying and campaign contribution activities go well beyond Senator Hatch. According to a 2015 article by Melanie Zanona for the *CQ Weekly*,[16] dietary supplement trade organizations spent $4.1 million on lobbying in 2014. And the industry's

[15] Lipton, E. *Support is Mutual for Senator and Utah Industry*. New York Times. June 10, 2011.
[16] Zanona, M. *Supplements Industry Masters the Hill*. CQ Weekly. June 1, 2015: 16–23.

PACs spent $1.1 million in 2014 elections, 63% of which went to Republican candidates. Some of the most prominent dietary supplement trade organizations and PACs are listed in the box below. It is an interesting exercise to visit their websites and get a sense of their mission and goals.

The Natural Products Association

Alliance for Natural Health

The Council for Responsible Nutrition

The Consumer Health Products Association

Alliance for Natural Health

The American Herbal Products Association

United Natural Products Alliance

Another important Congressional group with a very strong influence on legislative activities that affect the dietary supplement industry is the Dietary Supplement Caucus. This congressional caucus is made up of both Republican and Democratic lawmakers from both the Senate and the House of Representatives, many of whom serve on committees that have control over various aspects of dietary supplements and their regulation, or are from states with significant dietary supplement industry presence. The Natural Products Association provides an interesting description of the Congressional Dietary Supplement Caucus:

The DSC was founded in 2006 and provides a forum for the exchange of ideas and information on dietary supplements. The DSC serves as a bipartisan and bicameral group of legislators who facilitate discussions among lawmakers about the benefits of dietary supplements, provide tips and insights for better health and wellness, and promote research into the health care savings these products provide.

The caucus seeks to enhance Congressional attention to the role of dietary supplements in health promotion and address the regulation of the dietary supplement industry.

From: www.npainfo.org/App_Themes/NPA/docs/press/Position%20 Papers/Congressional%20Dietary%20Supplement%20Caucus.pdf

A Google search of the Dietary Supplement Caucus will yield a fascinating list of presentations to the caucus members and staff by various pro-supplement speakers. This may lead some health professionals and others to question the extent to which our legislators are receiving and understanding all sides of the science related to dietary supplement safety and effectiveness.

Given that Senator Hatch is a Republican and that a high proportion of the industry's campaign contributions are targeted to Republican candidates, it may be tempting to think of this as a partisan issue. After all, Republicans are stereotypically strong advocates for deregulation and minimal interference by the federal government in the free enterprise system. These types of policies and political agendas certainly play a significant role in the long-standing congressional opposition to expanded FDA regulatory authority of the dietary supplement industry. But this is not as simple as partisan politics might suggest. As is apparent by the membership of the Dietary Supplement Caucus, many Democrats are also opposed to expanded FDA oversight of this industry, albeit for often much different reasons. While some Democratic legislators, like their Republican colleagues, are responding to the influences of the industry lobbyists and the presence of dietary supplement industries in their home states and congressional districts, others are responding to a general public mistrust of drug company motives and profits and a general suspicion of the FDA and its relationships with drug companies and the traditional medical establishment. Many citizens who would describe themselves politically as Democrats or liberal are threatened by what they perceive to be the federal government interfering in their right to choose a more "natural" nontraditional approach to managing their health. They see unrestricted access to dietary supplements as an important part of this personal health strategy.

So the FDA is faced with the daunting challenge of dealing with a Congress that is heavily influenced by both the dietary supplement industry, and a general public constituency that wants minimal government interference in their access to dietary supplements. The one thing that the FDA does have on its side is science. And the science is pretty clear. The vast majority of dietary supplements on the market offer no significant health benefits and some pose significant health risks. Why doesn't the science prevail, particularly with the general public, most of whom presumably have no stake in the failure or success of the industry (unlike many members of Congress)? As discussed earlier in this chapter, the biggest users of dietary supplements are those with a college education or more. One would think that this demographic would be most likely to understand and accept the general scientific consensus regarding the general lack of health benefits from most dietary supplements. Unfortunately, the combination of slick marketing and a regulatory environment that gives the dietary supplement industry wide latitude regarding the health claims that can be made or inferred with little if any scientific proof overwhelms even some of the most educated, skeptical, and discerning consumers. This paradox between scientific consensus and personal belief and choice is not unique to the dietary supplement controversy. For example, how often have you heard individuals ridicule climate change "deniers" for not accepting the overwhelming scientific consensus linking global warming to human influence? And yet, how often have we heard some of these same individuals deny the overwhelming scientific consensus rejecting a link between vaccines and autism? Or arguing that Vitamin C is effective in curing the common cold, despite decades of research that has failed to demonstrate such an effect? When it comes to personal health, Americans are too often tempted by the easy cure … just take a pill. Or even better, just take a pill that is made from all natural herbal ingredients!

Thus, the regulation of dietary supplements is clearly a very complicated issue. For over 70 years, the FDA has tried to balance its responsibility to protect the public from dangerous or fraudulent health products, while preserving the public's right to free choice and access. In the subsequent chapters of this book, we will retrace the FDA's long and often frustrating attempts to navigate this balance. But first it is important that you have a basic understanding of how the federal government works and the roles of the various agencies responsible for the regulation of dietary supplements.

chapter three

Our federal food regulatory structure

It's time for a short refresher course in civics and the structure and function of our federal government. Throughout this book, we will be referring to many functions and responsibilities of the three branches of the government. We will also be referencing and discussing many different government documents. To fully understand the later chapters in this book, it is important for the reader to be familiar with how our legislative, executive enforcement, and judicial branches of government work. I will be focusing primarily on those aspects of our government structure and function that are most relevant to the topic of food regulation.

The U.S. Constitution establishes three branches to our federal government: the legislative branch, the executive branch, and the judicial branch. The Constitution also specifies the responsibilities of each branch. Many of these responsibilities are not particularly relevant to our discussion of food regulation. For example, among the responsibilities of the legislative branch is the power to declare war, the authority to approve treaties, the authority to coin money, etc. Clearly, these are not directly relevant to the regulation of our food supply, so we will not review them here. If you are interested in learning more about the broader constitutional structure and authority of our federal government, you may want to visit the National Archives, America's Founding Documents website at: www.archives.gov/founding-docs.

The Legislative Branch. *All legislative Powers herein granted shall be vested in a Congress of the United States, which shall consist of a Senate and House of Representatives.* (Article I, Section I of the U.S. Constitution).

Certainly, one of the most important responsibilities of the legislative branch (and the one most relevant to the regulation of our food supply) is the authority to write, debate, and pass bills. In order for a bill to become law, it must be voted upon favorably by a majority of members of both houses of Congress, and it must then be signed by the President. This is often a very slow and tedious process. Many bills never even make it to a vote in either house. They may "die in committee" (Congressional committee fails to vote on a bill) or be "tabled" (suspend consideration of a bill indefinitely). Sometimes, the two houses of Congress cannot agree on a mutually acceptable form of a bill. Or the President may veto the bill, despite it having been passed by the Congress. Overwriting a Presidential veto requires two-thirds vote of both houses of Congress; something that is often very difficult to accomplish.

So it should come as no surprise that most bills that are introduced in Congress never become laws. In fact you may find the actual numbers quite surprising. For example, for the 113th Congress (in session for the 2-year period between January 2013 and January 2015), approximately 10,000 pieces of legislation were introduced in Congress and only 296 laws were ultimately enacted.[1] There are many reasons for this low percentage of introduced bills that became law. For example, legislators often introduce bills in Congress to garner support from their constituents, yet knowing that they have no chance of actually becoming law. In addition, the intentionally slow and deliberate nature of the legislative process weeds out many of the bills that should rightfully not become laws, for any number of reasons. Considering how difficult and time-consuming it is for a new law to be enacted, it is important that those laws that are ultimately enacted be able to "stand the test of time." The U.S. Congress does not have the time, resources, nor desire to constantly go back and fix or repeal broken, weak, or outdated laws. Of course, that's not to say that it doesn't occasionally do this. Even the best laws need to be amended or even repealed at times. Yet, many laws passed by Congress in the 1800s are still important and relevant today, albeit perhaps with some amendments over the years to reflect changes in societal and cultural norms and priorities. One way that Congress can ensure that laws withstand the tests of time is to ensure that the wording of the law conveys the basic intent of what is to be required (or prohibited), while avoiding wording related to details and specifics that are likely to change over the years as new information related to the law becomes available. Our federal food laws are excellent examples of this.

In 1938, Congress passed and President Franklin D. Roosevelt signed into law the Food, Drug, and Cosmetic Act, often simply abbreviated as the FD&C Act of 1938. This is the primary federal food law that we still operate under today, although it has been amended many times over the past 80 years. This law, as amended, impacts the daily lives of every American every day, perhaps more than any other piece of federal legislation ever passed by Congress. The law remains powerful and relevant today, 80 years since its enactment, despite the fact that it deals with a topic (our food

[1] www.govtrack.us/congress/bills/statistics.

supply) that is constantly changing. To better illustrate this, let's briefly consider an actual example from the FD&C Act. The verbatim wording of one example section of the law, as written by Congress in 1938 states that

> **21 U.S. code (USC) § 331**
> *The following acts and the causing thereof are prohibited:*
> *(a) The introduction or delivery for introduction into interstate commerce of any food, drug, device, tobacco product, or cosmetic that is adulterated or misbranded.*

This section seems rather straightforward and clear. Congress is essentially stating that it is illegal to sell a food product (or drug, device, tobacco, or cosmetic) that is adulterated or misbranded. Of course, that begs the obvious question, when is a food considered adulterated or misbranded? For example, let's just consider food adulteration. Congress described in later sections of the law some of the conditions that would render a food "adulterated." Here is an example from the original 1938 law.

> **21 USC §342**
> *Adulterated food*
> *A food shall be deemed to be adulterated*
> *(4) if it has been prepared, packed, or held under insanitary conditions, whereby it may have become contaminated with filth, or whereby it may have been rendered injurious to health;*

These two quoted sections of the FD&C Act have not changed since the original passage of the law in 1938. Note that when Congress wrote these sections of the law, they did not specify what should be considered "insanitary conditions," or what should be considered "filth," or what should be considered "injurious to health." Congress likely anticipated that the standards for these conditions may change over the years, as new knowledge regarding the science of food safety and health is acquired and techniques for detecting and monitoring the safety of our food supply improve. The Legislative Branch expects that the Executive Branch, charged with enforcing the laws, will provide the necessary details needed to keep the law up to date with the latest scientific information and with the ever-evolving public expectations regarding the safety of their food supply, which brings us to a brief discussion of the Executive Branch of the federal government.

The Executive Branch: *The executive power shall be vested in a President of the United States of America.* (Article II, Section I of the U.S. Constitution).

...; he shall take Care that the Laws be faithfully executed (Article II, Section 3 of the U.S. Constitution).

As with the Legislative Branch of the government, the U.S. Constitution specifies many responsibilities of the Executive Branch. But its responsibility to "take care that the laws be faithfully executed" is the one responsibility most relevant to food regulation.

When many of us think of the Executive Branch, we think of the President and the White House. Based on that impression, we may be tempted to think of the Executive Branch as the smallest of the three branches of the federal government. In fact, it is by far the largest component of the federal government. The Legislative and Judicial branches of the federal government combined employ approximately 60,000 people. In contrast, the Executive Branch employs over two million civilians, and that figure is excluding uniformed military personnel.[2] Clearly, the Executive Branch consists of much more than the President and the White House. The immediate staff of the White House, known as the Executive Office of the President, does employ over 2,000 people, obviously not all of whom are physically located in the White House. However, the vast majority of Executive Branch employees are associated with cabinet-level departments and independent (non-cabinet-level) agencies. There are currently 15 cabinet-level departments in the Executive Branch (listed in the box below).

Department of Agriculture	Department of the Interior
Department of Commerce	Department of Justice
Department of Defense	Department of Labor
Department of Education	Department of State
Department of Energy	Department of Transportation
Department of Health and Human Services	Department of the Treasury
Department of Homeland Security	Department of Veterans Affairs
Department of Housing and Urban Development	

[2] www.opm.gov/policy-data-oversight/data-analysis-documentation/federal-employment-reports/historical-tables/executive-branch-civilian-employment-since-1940/.

In addition, there are more than 50 independent agencies in the Executive Branch; far too many to list all of them here. But here is a partial list to give you some sense of where some of the more recognizable of these agencies fit into our federal government structure.

Central Intelligence Agency	National Science Foundation
Consumer Product Safety Commission	National Transportation Safety Board
Environmental Protection Agency (EPA)	Nuclear Regulatory Commission
Federal Communications Commission	Occupational Safety and Health Review Commission
Federal Elections Commission	Peace Corps
Federal Reserve System	Securities and Exchange Commission
Federal Trade Commission (FTC)	Small Business Administration
National Aeronautics and Space Administration	Social Security Administration

You should now have a good appreciation for the immense size of the Executive Branch. However, only a few of these departments and agencies have food regulatory responsibilities, and still fewer have authority over the regulation of dietary supplements. But let's start with a quick overview of the roles of the various departments and agencies that have some level of food regulatory authority, and then we will focus in detail on those with particular authority related to dietary supplements.

The U.S. Customs Service and Border Patrol: This organization is part of the Department of Homeland Security. Among its food-related regulatory responsibilities is the enforcement of imported food laws and regulations. It also oversees the re-exportation or disposal of imported foods that are not in compliance with U.S. laws/regulations.

The Alcohol and Tobacco Tax and Trade Bureau: This bureau is part of the Department of the Treasury. Its primary food-related responsibility is the regulation of the labeling of alcoholic beverage.

The Centers for Disease Control and Prevention: The Centers for Disease Control and Prevention (CDC) is part of the Department of Health and Human Services. Among its many functions, the CDC is responsible for tracking and investigating the causes of food-borne illness in this country.

The National Institutes of Health: The National Institutes of Health (NIH) is also part of the Department of Health and Human Services. It consists of 27 institutes and centers whose primary functions are bio-medical research and education. Although these institutes and centers that make up the NIH do not have direct responsibilities related to food regulation, they have at times played very important indirect roles in the regulation of dietary supplements, and they continue to do so today. For example, as we will see later in this book, the National Cancer Institute was closely connected to the early controversies related to health claims on the labeling and advertising of foods in the 1980s. And the NIH's National Center for Complementary and Integrative Health is actively involved in research and education related to alternative thera-pies, many of which involve herbal and other dietary supplement prod-ucts. In addition, Congress mandated the establishment of an Office of Dietary Supplements when it passed the Dietary Supplement Health and Education Act (DSHEA) in 1994. This office within the NIH is responsible for supporting research, scientific evaluation, and education related to dietary supplements.

The Environmental Protection Agency: As indicated in the box earlier, the EPA is an independent agency within the Executive Branch (not part of a cabinet-level department). It has several food-related responsibilities, including the approval of new pesticides and the establishment of accept-able levels of pesticide residues on food.

The Food Safety and Inspection Service: Food Safety and Inspection Service is part of the Department of Agriculture. It has regulatory authority over meat and poultry (and meat and poultry products) and egg product production and the inspection of facilities that produce these products.

The Federal Trade Commission: The FTC is also an independent agency within the Executive Branch. Among its responsibilities related to food is the authority to regulate food advertising and the prevention of "unfair or deceptive" practices in food advertising. As such, it plays a very impor-tant role in the marketing and advertising of dietary supplements. We will be considering the role of the FTC in this capacity in more detail later in this book.

The Food and Drug Administration: The U.S. Food and Drug Administration (FDA) is part of the Department of Health and Human Services. It has, by far, the broadest authority over food regulation in the United States (as well as food imports). It is primarily responsible for the enforcement of the FD&C Act (mentioned earlier in our discussion of the Legislative Branch). This includes all of the amendments to the FD&C Act since its first enactment in 1938. Several of these amendments directly impact the regulation of dietary supplements. Most notable in this regard is the DSHEA. This amendment to the FD&C Act was signed into law in 1994 and will be the focus of much discussion and consideration throughout this book.

How does the Executive Branch carry out its constitutional mandate to *"take care that the laws be faithfully executed"*? The Executive Branch department or agency charged with enforcing the particular law will write detailed regulations specifying how the law will be enforced. This is officially known as the "Rulemaking Process." A basic understanding of the Rulemaking Process will be helpful as we move into further discussions of the FDA's historical regulation of the dietary supplement industry. But before we review this process, we need to first briefly explain the purpose and differences between some important and relevant government documents.

The U.S. Code: In our earlier overview of the Legislative Branch of the government, I quoted a couple of short sections of the FD&C Act. One of the quotes is presented again here:

21 USC § 331
The following acts and the causing thereof are prohibited:
(a) The introduction or delivery for introduction into interstate commerce of any food, drug, device, tobacco product, or cosmetic that is adulterated or misbranded.

Notice the citation that I included at the beginning of this quoted section of the law (21 USC § 331). The "USC" stands for the "United States Code." The USC is a government publication of all of the laws currently in effect, organized (or codified) by subject matter or topic. The number "21" preceding the USC in the citation refers to a particular title of the USC. There are 53 titles to the USC, each representing a different subject matter area. Title 21 of the USC covers laws related to "Food and Drugs." Thus, the FD&C Act and its amendments are codified and published in Title 21. The "§ 331" in the earlier citation refers to the particular "part" of Title 21 of the USC. There are a number of very good online links to the USC and other government documents. One that I particularly like is the U.S. Government Publishing Office, Federal Digital System at www.gpo.gov/fdsys/. You might find it helpful to take a quick look at this site and browse some of the titles to the USC.

The Federal Register: The Federal Register is a daily (weekday) publication of all activities of the Executive Branch. It contains many different categories of information, including Presidential proclamations, notices of meetings and hearings, and details related to new and proposed regulations. You can access an online version of the Federal Register through the www.gpo.gov/fdsys link mentioned earlier or through its own link at www.federalregister.gov/. This link is particularly helpful and user-friendly. An example of a Federal Register citation would be *72 FR 62149*, where 72 refers to the volume number (with each year assigned a consecutive number), FR refers to the Federal Register, and 62149 refers to the starting page number in volume 72. Page numbers are cumulative for the entire year (volume number). So it is possible to have as many as 100,000 pages or more in one volume (year) of the Federal Register.

The Code of Federal Regulations: The Code of Federal Regulations, or CFR, is a publication of all regulations currently in effect organized by subject matter (title). There are 50 titles to the CFR, and most of them coincide with the titles in the USC. For example, Title 21 of the CFR covers regulations related to "Food and Drugs" and enforce the laws that are covered in Title 21 (Food and Drugs) of the USC. You can access the CFR online from the www.gpo.gov/fdsy site or through a dedicated Government Publishing Office link to the eCFR at www.eCFR.gov.

There are many other government documents that are important to food regulation, but these three are certainly the most relevant to our discussion of the topic. Now that you have a basic understanding of the function of these government documents, we can move on to a review of the Executive Branch's Rulemaking Process.

When Congress writes a new law, they typically will specify which particular department or agency within the Executive Branch will be responsible for enforcing the law. This is not always the case. Occasionally, they will leave it up to the President to decide. But if you were to read through the FD&C Act in Title 21 of the USC, you would quickly notice that throughout the wording of the law, it frequently refers to "the Secretary." In fact, 21 USC § 321(d) specifically states, *'The term "Secretary" means the Secretary of Health and Human Services.'* So Congress made it very clear that the Department of Health and Human Services within the Executive Branch is responsible for enforcing this law. Notice that Congress did not specify that the FDA should enforce the law. However, the Secretary of Health and Human Services has the authority to delegate that responsibility to the FDA.

Recall in our earlier discussion of the legislative process that laws are typically written succinctly, with a focus on the overarching goals and intent of the law, while avoiding too much focus on details that are better handled by the Executive Branch enforcement agencies. As mentioned earlier, the Executive Branch provides these enforcement details by writing regulations. Regulations have the power of law. That is, if you violate a regulation, you are violating the law that the particular regulation is enforcing. The procedure for writing regulations is known as the *Rulemaking Process*. The www.federalregister.gov website has a very nice explanation of the Rulemaking Process.[3] I am going to briefly summarize the description of the process here, but for a more in-depth explanation, you may want to visit their website.

One of the most important features of the Rulemaking Process is the opportunity, or more precisely, the requirement for public input in the process. Certainly, that is true to some extent in the legislative process as well. Often, there are public hearings when Congress is considering a bill, and you always have the right and opportunity to contact your Congressperson to express your opinion on a bill or other legislative action. However, the Executive Branch Rulemaking Process specifically mandates public input consideration. This public input process can start even before an agency has formally proposed a new regulation. Very often in the early stages of the Rulemaking Process, the agency will publish what is known as an "Advance Notice of Proposed Rulemaking," or ANPRM, in the Federal Register. The purpose of this ANPRM is to gather information that may help the agency in writing the regulation, or perhaps, even whether to write it at all. The agency can "test the waters" so to speak, to see if opposition to a possible new regulation is too strong, for example. The publication of the ANPRM in the Federal Register will typically include extensive background information on the particular topic, the agencies reasoning and justification for a possible new regulation, and detailed information on how the public and other interested parties can provide input.

Based on the input from the ANPRM, the agency may proceed to formally publish a proposed rule, or Notice of Proposed Rulemaking (NPRM), in the Federal Register. The format for this announcement is very similar in structure to the "Proposed Rule" illustrated in Figure 3.1. As with the ANPRM, the proposed rule would allow a period for public comment. This period is sometimes extended if the agency feels it needs more information or if it feels that the public needs more time to comment.

Following the comment period for the proposed rule, the agency may proceed to publish the Final Rule. It is important to note that unlike the legislative process, which requires voting in both houses of Congress, there is no voting process in Executive Branch rulemaking. The agency will simply consider all comments, scientific data, and expert opinions in deciding on whether or not to proceed to a Final Rule and the specific details of that Final Rule. The Final Rule announcement in the Federal

---

[3] www.federalregister.gov/uploads/2011/01/the_rulemaking_process.pdf.

DEPARTMENT OF HEALTH AND
HUMAN SERVICES

Food and Drug Administration

21 CFR Part 101

[Docket no. FDA-2012-N-1210]

RIN 0910-AF22

Food Labeling: Revision of the
Nutrition and Supplement Facts Labels

AGENCY: Food and Drug Administration

ACTION: Proposed Rule.

SUMMARY: The Food and Drug
Administration (FDA, the Agency, or
we) is proposing to amend its labeling
regulations for conventional foods
and dietary supplements to provide
updated nutrition information on the
label to assist consumers in maintaining
healthy dietary practices. The updated
information is consistent with.............

DATES: Submit either electronic or.......

ADDRESSES: You may submit comments,
identified by Docket.......

FOR FURTHER INFORMATION CONTACT:
Blakeley Fitzpatrick, Center for Food
Safety and Applied Nutrition ....

BACKGROUND: etc

Figure 3.1 Sample format for a typical Federal Register "Proposed Rule" announcement.

Register is again very similar in format to the ANPRM and NPRM notices. Every Final Rule in the Federal Register will include an "Effective Date." This is the date on which the new regulation will go into effect. This can be soon after the announcement in the Federal Register, or several years beyond the announcement date in situations where parties affected or impacted by the new regulation(s) need time to prepare for the changes that the regulation requires. The Final Rule announcement in the Federal Register will also include the exact wording of the regulation(s) as it will appear in the CFR once they become effective.

There is another very important component of the Rulemaking Process. When proposing new regulations, the Executive Branch agencies must identify their legal authority for establishing the new regulation. For example, the FD&C Act establishes limits on what the FDA can do in its efforts to enforce the law. I think it is fair to generalize that it is not unusual for an Executive Branch agency, such as the FDA, to push the limits of this authority at times. This will inevitably lead to occasional disputes between the Executive Branch agency and those who may be adversely affected by the agency's regulatory authority. How are these disputes settled? That brings us to the third branch of the federal government.

The Judicial Branch. *The judicial power of the United States shall be vested in one Supreme Court, and in such inferior courts as the Congress may from time to time ordain and establish* (Article III, Section I of the U.S. Constitution).

The Judicial Branch of the federal government consists of a federal court system with three levels: the district court, the circuit court, and the Supreme Court. All levels of the federal courts decide cases that involve issues related to the U.S. Constitution or cases involving federal laws. There are 94 district courts in the United States, and there are 13 circuit courts (United States Court of Appeals). As you know, there is only one U.S. Supreme Court. The district courts are trial courts with decisions made by a jury. The circuit courts and the Supreme Court do not use juries. Rather, decisions are made by the circuit court judges or the Supreme Court Justices. Federal criminal and civil cases typically begin within the district court system. Decisions of the district court system can then be appealed to the circuit court of appeal. Decisions of the circuit court may then be appealed to the U.S. Supreme Court. Decisions of the Supreme Court are final and can only be altered by a subsequent Supreme Court decision or by a constitutional amendment.

So, let's consider how this federal court system may be involved in food regulation issues. One obvious example would be a situation where an individual or a company violates some provision of the FD&C Act (or its associated regulations). The FDA could then file criminal charges against the company in U.S. District Court. Here is an example from an FDA press

release from April 2015 that describes a criminal charge brought against a
dietary supplement manufacturer:

April 15, 2015: Owner of Harrisburg Diet Supplement Business Charged with Selling Misbranded Drugs

f SHARE   ⚇ TWEET   in LINKEDIN   ⦿ PIN IT   ✉ EMAIL   🖨 PRINT

**Food and Drug Administration
Office of Criminal Investigations**

U.S. Department of Justice Press Release

For Immediate Release   **United States Department of Justice**
April 15, 2015   **Middle District of Pennsylvania**

The U.S. Attorney's Office of the Middle District of Pennsylvania announced today that a criminal information was
filed in U.S. District Court in Harrisburg, Pennsylvania, charging Cheryl Floyd, 52, Harrisburg, owner of Floyd
Nutrition LLC, with introducing misbranded drugs into interstate commerce and money laundering.

According to U.S. Attorney Peter Smith of the Middle District of Pennsylvania, Floyd, aka Cheryl Floyd Brown, is
owner and operator of an internet-based business known as Floyd Nutrition LLC, based at her Harrisburg residence
and warehouse facilities in the Harrisburg area.

The items offered for sale between 2010 and 2014 were allegedly purported all-natural dietary supplements sold as
weight loss products. They allegedly contain the drugs sibutramine and phenolphphthalein which are not listed as
ingredients in the product labels.

According to U.S. Food and Drug Administration (FDA), sibutramine was the active pharmaceutical ingredient in
Meridia, a prescription weight loss drug removed from the market in 2010 following studies that showed increased
heart attack and stroke in the studied population. Phenolphphthalein was an over-the-county drug until 1999 when

But beyond federal criminal cases, the federal court system is also
often involved in settling disputes related to whether or not a law passed
by Congress is in conflict with the U.S. Constitution, or whether a regulation established by the Executive Branch is in violation of the U.S.
Constitution or of federal law. This process is known as "judicial review,"
and it is an essential part of the "checks and balances" functionality of
the federal government. The "checks and balances" system is designed
to ensure that no branch of the federal government abuses or exceeds its
power and authority. There are many "checks and balances" functions

built into our federal government structure. For example, the Judicial Branch has the power of judicial review of Executive Branch regulations and executive orders, while the Executive Branch has the power to appoint federal judges, including Supreme Court justices. And the Judicial Branch has the power of judicial review of Legislative Branch laws, while the Legislative Branch has the power to approve federal judge nominations and, if necessary, impeach federal judges.

So, if an individual, company, or special interest group feels that the FDA has exceeded its authority in establishing a new regulation, that individual, company, or special interest group can take the FDA to court, potentially all the way to the U.S. Supreme Court, to try and halt the new regulation. As we shall see in later chapters of this book, these challenges have played a very important role in the history of dietary supplement regulation.

chapter four

The early history of dietary supplement regulation

In earlier chapters of this book, I mentioned the Dietary Supplement Health and Education Act (DSHEA). Although we have not yet discussed the specific provisions of this law, it may already be clear that passage of this law in 1994 radically altered the regulation of dietary supplements in the United States. The DSHEA sets strict limits on the regulatory role that the federal government can play in the manufacture, labeling, and advertising of these products. It has now been more than two decades since this law was passed, and although there have been a few minor refinements to the law over these years, it remains essentially unchanged from when it was first passed. Looking back on these past two decades, I think it would be fair to say that dietary supplement companies have been generally pleased with the protections that the DSHEA has provided the industry. However, the Food and Drug Administration (FDA) has at times been frustrated by this law's restrictions on the agency's authority to protect the public from what it considers to be some potentially dangerous products or products that make misleading or exaggerated claims regarding health benefits. In Chapter two, we examined some of the reasons why people take dietary supplements. We also briefly discussed the influences that this multibillion dollar industry has on the legislative branch. These factors certainly played a significant role in the passage of DSHEA in 1994. But perhaps, even more notable is the role that the FDA may have inadvertently played in the passage of this law as an indirect result of some of its actions in the preceding decades. In this chapter, I will offer a very brief history of the early years of federal food and drug regulation in the United States and set the stage for our later focus on dietary supplement regulation and the events leading up to the DSHEA.

It might seem surprising to us today, but 150 years ago, the federal government essentially had no control over the manufacture and sale of either food or drug products in the United States. In fact, the prevailing legal doctrine overseeing the sale of these products was *caveat emptor* or "let the buyer beware." Before the industrial revolution, the application of this doctrine made some sense. At that time, most people produced their food themselves, or had close personal contact with individuals from whom they bought their food. This was true of most medicinal products as well. If there was a problem with a product that a person purchased from a local producer or merchant, that producer or merchant would hear about it personally from the buyer. This could result in an immediate negative economic consequence to the seller (i.e., everyone in that community would stop buying from that particular seller). Other factors also likely contributed to the persistence of the *caveat emptor* doctrine at that time. For example, advertising and labeling claims were typically considered "free speech" and protected by the First Amendment of the Constitution, even if the advertising or labeling was false or misleading. Additionally, there was essentially no effective local, state, or federal law enforcement authority empowered to crack down on products that were dangerous or fraudulent. And the fourth amendment to the Constitution protected citizens from unlawful search and seizures (without a warrant). This protection was logically extended to privately owned food or drug manufacturers, limiting the ability of government agents from inspecting manufacturing facilities without a search warrant.

But beginning in the latter part of the 19th century, the doctrine of *caveat emptor* began to shift to a doctrine of *caveat venditor*: "let the seller beware." This shift was driven largely by the industrial revolution's urbanization of America, resulting in the buyer often being far removed from personal contact with the manufacturer. Unscrupulous food or drug manufacturers could adulterate or mislabel their products with little fear that the buyer, who now may live very far away from the manufacturer, would have any recourse. By the late 1800s, as the incidences of food and drug adulteration and mislabeling continued to grow, the citizens began to expect more protection from their government. Championing this effort was the chief chemist for what was then the Bureau of Chemistry in the U.S. Department of Agriculture (USDA), Dr. Harvey W. Wiley. Dr. Wiley's early interest in food adulteration was focused on detecting adulteration of sugars and syrups for the Indiana State Board of Health while he was a professor at Purdue University. But after becoming Chief Chemist for the USDA Bureau of Chemistry, the U.S. Congress funded his lab to study the safety of the food preservatives that were in common use at the time. Dr. Wiley had a reputation as a meticulous and precise chemist and scientist. It was these traits that led him to become the founding member of the Association of Official Analytical Chemists, an organization that serves to this day to provide scientists with standardized methods of chemical and microbiological analysis. He approached his analysis of food adulteration with the same zeal and dedication, even employing a group of human volunteers to systematically examine and record the effects of food adulterants and preservatives on human health. This group of volunteers became popularly known as "The Poison Squad." Of course, this would be completely unethical to do today. But at the time, Dr. Wiley and his "Poison Squad" helped draw national attention to the dangers lurking in everyday foods routinely sold to unsuspecting consumers. And Dr. Wiley's talents went far beyond his scientific and laboratory skills. He was a tireless advocate for passage of federal legislation to address these health and economic dangers in the food supply, and he used his excellent public speaking skills and powers of persuasion to rally public support, enlist the press in his mission, and lobby the Congress to enact legislation.

Dr. Harvey W. Wiley (from USA.gov image archive).

The Poison Squad at work (from USA.gov image archive).

While few would dispute the enormous impact of Dr. Wiley's efforts, it may have been the publication of a novel that was the final impetus for federal legislative action. In 1906, Upton Sinclair published his book, *The Jungle*, describing the horrendous working, health, and food safety conditions in meat-packing plants in Chicago (see short excerpt in box later).

The public was horrified and nauseated by what the author described, and President Theodore Roosevelt immediately commissioned an investigation of the meat-packing industry. After just a few weeks, the commission confirmed the conditions described in *The Jungle*, and the President pressed the Congress to act immediately, resulting in the passage of the Pure Food and Drug Act on June 30, 1906. This first comprehensive federal food and drug law accomplished some very important goals, including defining when a food is adulterated or misbranded, allowing for seizure of adulterated or misbranded food, and allowing for fines or imprisonment for violators of the law. But it also had some important shortcomings. For example, it neither provided for factory inspections nor established prosecutorial standards for when a food could be declared adulterated. The law also required proof of intent to adulterate or misbrand a food. Thus, the defendants could simply argue that they were not aware that they had violated the law. And perhaps most importantly, Congress did not appropriate sufficient funds to adequately enforce the new law.

There was never the least attention paid to what was cut up for sausage; there would come all the way back from Europe old sausage that had been rejected, and that was moldy and white – it would be dosed with borax and glycerin, and dumped into the hoppers, and made over again for home consumption. There would be meat that had tumbled out on the floor; in the dirt and sawdust, where workers had tramped and spit uncounted billions of consumption germs. There would be meat stored in great piles in rooms; and the water from leaky roofs would drip over it, and thousands of rats would race about it. It was too dark in theses storage places to see well, but a man could run his hand over these piles and sweep off handfuls of the dung of rats. These rats were nuisances, and the packers would put poison bread out for them, they would die, and then rats, bread, and meat would go into the hoppers together.

Excerpt from *The Jungle* by Upton Sinclair

In the years following the passage of the 1906 law, a number of important Congressional amendments, court rulings, scientific advances, and executive branch reorganizations occurred. First, with regard to the reorganization of the Bureau of Chemistry, in 1927 it was split, resulting in establishment of a regulatory agency known as the Food, Drug, and Insecticide

Administration, but still located within the USDA. In 1930, the agency name was shortened to the Food and Drug Administration. And then in 1940, the FDA was moved from the USDA to what was then known as the Federal Security Agency. The Federal Security Agency became the Department of Health, Education and Welfare in 1953, and then the Department of Health and Human Services in 1979, where it remains today.

Some of the important amendments to the 1906 law and relevant court rulings are listed in the timeline box later.

1911: In **U.S. v. Johnson,** the Supreme Court rules that the 1906 Pure Food and Drug Act does not prohibit false therapeutic claims but only false and misleading statements about the ingredients or identity of a drug.

1912: Congress enacts the **Sherley Amendment** to overcome the ruling in U.S. v. Johnson. It prohibits labeling medicines with false therapeutic claims intended to defraud the purchaser, a standard difficult to prove.

1913: Gould Amendment requires that food package contents be "plainly and conspicuously marked on the outside of the package in terms of weight, measure, or numerical count."

1914: In **U.S. v. Lexington Mill and Elevator Company,** the Supreme Court issues its first ruling on food additives. It ruled that in order for bleached flour with nitrite residues to be banned from foods, the government must show a relationship between the chemical additive and the harm it allegedly caused in humans. The court also noted that the mere presence of such an ingredient was not sufficient to render the food illegal.

1924: In **U.S. v. 95 Barrels Alleged Apple Cider Vinegar,** the Supreme Court rules that the Food and Drugs Act condemns every statement, design, or device on a product's label that may mislead or deceive, even if technically true.

Source: About FDA: Significant Dates in U.S Food and Drug Law

It is clear from this timeline that all three branches of the federal government (Congress, the courts, and the Executive Branch enforcement agencies) were already focused on problems with the 1906 law as it related to false or misleading labeling in general, and in particular, false therapeutic claims on product labels. With regard to therapeutic claims, it would appear from the timeline that these were generally associated with drug products. However, the Bureau of Chemistry was also dealing with therapeutic claims on food products. In these early years following enactment of the 1906 law, the Bureau of Chemistry was frequently faced with regulating therapeutic claims on products that fell somewhere in between the legal definition at the time for a food versus a drug. Peter Barton Hutt published an excellent historical review of the government's regulation of health claims in food labeling and advertising.[1] He points out that when enforcing the 1906 law, the government did not find it necessary to make a determination as to whether the product was a food or a drug, only that the product's label was "false or misleading." There were many products on the market at the time that blurred the distinction between food and drug. Cod liver oil, for example, was a food item that was often prescribed by physicians for its medicinal properties. Hence, it could be considered as both a food and a drug, depending on its intended use.

It was also around this time that the science of vitamins and their role in nutrition and health was emerging. In fact, the term "vitamine" was first coined in 1912 by Casimir Funk, a Polish biochemist, to describe unidentified "vital amine" compounds that were recognized to be essential for normal growth and health in animal studies.[2] When it was subsequently learned that many of these compounds were not chemically "amines" (specific nitrogen-containing compounds), the "e" was dropped from the word, leaving the term "vitamin" that we know today. Over the next several decades, many of these individual vitamins were identified, structurally characterized, and, when not supplied in adequate amounts in the diet, linked to specific deficiency diseases.

The general public was understandably intrigued by the discoveries of these essential dietary factors. Inevitably, this led some unscrupulous food and drug manufacturers to tout exaggerated therapeutic claims for these newly identified vitamins on their product labels and in product advertising. In fact, nearly 100 years ago, the then "Food, Drug, and Insecticide Administration" was already recognizing a problem with the regulation of some of these products; products that would later come to be known as "dietary supplements." In Peter Barton Hutt's review, he quotes from a 1929 Annual Report of the Food, Drug, and Insecticide Administration, which clearly expresses the government's concerns and foretells of the upcoming century long struggle between the FDA and the dietary supplement industry.

---

[1] Hutt, PB. Government Regulation of Health Claims in Food Labeling and Advertising. *Food, Drug, Cosmet. Law J.*, 41: 3–73 (1986).
[2] Rosenfeld, L. Vitamine – Vitamin. The Early Years of Discovery. *Clin. Chem.*, 43(4): 680–685 (1997).

Some manufacturers have taken advantage of the general public interest in vitamins to exploit preparations as sources of vitamins when the facts do not warrant it. The administration has attempted to put a damper on unwarranted labelings of this character, seeking as far as possible to induce the manufactures to make their own investigations and limit the statements on their labels to demonstrable facts.

Source: 1929 Annual Report. Federal food, drug, and cosmetic law: administrative reports, 1907–1949. Drug Law Institute (US). 1951. Commerce Clearing House, Chicago.

It was not just the government that was concerned. Even in these early years of vitamin discovery and understanding, some scientists were already recognizing that there was no scientific basis for ingesting more than the small amounts of these substances necessary to prevent deficiency: levels that can be obtained by simply eating a healthy diet. In 1922, Dr. L. Emmett Holt, a renowned pediatrician, author, and faculty member of the Columbia University College of Physicians and Surgeons published an article in the *Journal of the American Medical Association* on the role of vitamins in health and disease.[3] In this article, he made some prophetic comments.

... it does not seem probable that ... improvement in general nutrition is likely to follow [vitamin] administration in larger amounts, when no real deficiency exists. In other words, benefits from an excess of any of them seems most improbable, and we lack proof that such is the case.

This is a remarkable statement for the time, considering how early it was in our understanding of the science and nutrition of vitamins. Yet today, nearly a century later, most nutritional scientists would conclude that, despite our much-improved knowledge of the role of vitamins and minerals in metabolism, health, and disease, Dr. Holt's statement is generally as true today as it was then.

[3] Holt, LE. The Practical Application of the Results of Vitamin Studies. *JAMA,* 79(2): 129–132 (1922).

Later in the paper, Dr. Holt goes on to comment:

> *It is most unfortunate that the popular interest in vitamins, which is the result of so much publicity as has been given to this whole subject, should be exploited commercially. ... The medical profession, at all events, should not be carried along in the popular current.*

So even as early as 1922, Dr. Holt was concerned about the proliferation of false and misleading claims regarding the unproven health benefits of vitamins that were already being foisted on a public eager for simple "one pill" solutions to many of the ailments of the time. But interestingly, on one point history would prove Dr. Holt quite wrong. Toward the end of the article, Dr. Holt writes

> *The use of vitamins without definite indications will be popular for a period but, like other fads, it will pass.*

One can only wonder what Dr. Holt would think about the proliferation and popularity of dietary supplements today and the kinds of health claims (implicit or explicit) that are being made to promote its sale.

chapter five

Dietary supplements
Foods or drugs?

The path to our current regulation framework for dietary supplements has been rather circuitous over the past 100 years. In tracing this route, it is helpful to examine it in the context of the history of the legal and popular cultural definitions of the terms "food" and "drug." In 2008, Lewis A. Grossman, Professor of Law at American University, published an article in the Cornell Law Review that did just that.[1] He notes that at the time of the passage of the 1906 Pure Food and Drug Act, food was legally defined as, "... *all articles used for food, drink, confectionary, or condiments by man or other animals, whether simple, mixed, or compound.*" This definition reflected the changing cultural perception of food at the time to include not only substances that provide the major macronutrients that support energy and growth (proteins, fats, and carbohydrates) but also substances whose primary function is related to taste and flavor (confectionary and condiments), as well as animal food. It is also worth noting that in the 1906 law Congress did not explicitly exclude drugs from the legal definition of food. It clearly recognized at the time that some foods could also be considered drugs, such as the cod liver oil example mentioned in the previous chapter.

The Pure Food and Drug Act of 1906 defined a "drug" as "... *all medicines and preparations recognized in the United States Pharmacopoeia (USP) or National Formulary (NF) for internal or external use, and any substance or mixture of substances intended to be used for the cure, mitigation, or prevention of disease of either man or animals.*" Again, it is noteworthy to consider this definition in a historical context. Prior to the early part of the 20th century, most "drugs" consisted of unrefined and unformulated plant-based products. These types of products could certainly be found in the USP. The USP, established in 1820 and published annually by the nonprofit USP Convention, is a compendium of drugs and drug information. Since it has been around for so long, many substances listed in the USP could satisfy both the legal and popular definitions of both foods and drugs. Many spices and herbs, for example, could meet either definition, depending

---

[1] Grossman, LA. Food, Drugs, and Droods: A Historical Consideration of Definitions and Categories in American Food and Drug Law. *Cornell Law Rev.*, 93: 1092–1148 (July 5, 2008).

on their intended use. But Congress in 1906 also recognized that science was increasingly moving toward drugs that were "formulated" by trained pharmacists from various natural or chemical constituents and that these types of products would be more typically and logically listed in a compendium such as the National Formulary (NF). Hence, they included the listing of a substance in either the USP or NF as a consideration in determining whether the substance met the legal definition of a drug. Perhaps most notable in the 1906 drug definition is the provision that a substance is legally a drug if it is "intended" to cure, mitigate, or prevent disease. In determining the "intended" use of the product, consideration could be given to many factors, including the labeling, advertising, and even promotion of the product.

So where did that leave "vitamin" products during this early period of federal food and drug regulation? That is not entirely clear. Recall that the term "vitamin" was not even coined until 1912. But according to the Food and Drug Administration historian John Swann,[2] despite the lack of explicit reference to vitamins or similar products in the 1906 law, the law (and the subsequent 1914 Federal Trade Commission Act that prohibited false or misleading advertising) was nevertheless reasonably effective at reducing the number of fraudulent labeling and advertising health or disease claims for food products, in general. And the Bureau of Chemistry (and later its successor FDA) was often successful in pursuing legal actions against many adulterated or misbranded products. In addition, in light of the restrictions imposed by these laws with regard to fraudulent and misleading health or disease claims, food manufacturers began to shift from promoting their products' health and disease-preventing attributes, to more emphasis on their taste and convenience attributes.

Still, as I mentioned in the previous chapter, the 1906 law did have some significant limitations. Various amendments and court decisions helped refine the law and address some of these limitations. But ultimately, it required an entirely new law to address the regulatory and enforcement requirements of a rapidly expanding food and drug industry in the United States. On June 25, 1938 President Franklin D. Roosevelt signed into law the Food, Drug, and Cosmetic Act (FD&C Act). This is the law that we still operate under today, although it has been amended many times over the past 80 years. Included in the language of the 1938 FD&C Act were new definitions for the terms "food" and "drug" that would come to have important implications to the dietary supplement industry and its regulation. The FD&C Act modified the legal definition of food just a bit, eliminating reference to confectionaries and condiments, as the

---

[2] Swann, JP. The History of Efforts to Regulate Dietary Supplements in the USA. *Drug Test. Anal.*, 8: 271–282 (2016).

1938 Congress apparently felt that these were now obviously foods, and it was no longer necessary to explicitly include them in the definition. Interestingly, Congress did feel it was necessary to specify gum as a food (and this still remains in the definition of food today).

> **21 U.S. code (USC) § 321(f)**
> The term "food" means (1) articles used for food or drink for man or other animals, (2) chewing gum, and (3) articles used for components of any such article.

This legal definition of the term "food" may seem a bit confusing at first glance. Congress is saying that the term "food" means "article used for food." Essentially, the law is referring to the common sense definitions of what is a food. In determining whether a substance is "used for food," we need to consider what we use food for. This is fairly simple to answer. We use food for taste, aroma, and nutritive value. Thus, if the primary purpose of a substance is to provide taste, aroma, and/or nutritive value, then it is a food. This interpretation of the definition will have important implications in the evolution of dietary supplement regulation.

The 1938 law also included a much more expanded definition of a "drug." This new definition would also come to have a significant impact on the regulation of dietary supplements in the years to follow.

> **21 USC § 321(g)**
> The term "drug" means (1) articles recognized in the official *United States Pharmacopoeia*, official Homeopathic Pharmacopeia of the United States, or official National Formulary, or any supplement to any of them; and (2) articles intended for use in the diagnosis, cure, mitigation, treatment, or prevention of disease in man or other animals; and (3) articles (other than food) intended to affect the structure or any function of the body of man or other animals; and (4) articles intended for use as a component of any article specified in clause (1), (2), or (3); but does not include devices or their components, parts, or accessories.

As you can see in the box above, the 1938 legal definition of a drug preserved the wording regarding articles included in various official compendia (part 1 of the definition), and it preserved the wording regarding articles "intended" to diagnose, cure, mitigate, treat, or prevent a disease. But it added a new third clause, "articles (other than food) intended to affect the structure or any function of the body..." This new consideration in the definition of a drug is worth looking at more closely. Notice that in part (2) of the drug definition, Congress recognized that an article could be legally considered a food and still satisfy this legal definition of a drug (intended to treat a disease). In other words, according to part (2) of this drug definition, the food and drug definitions are not mutually exclusive. Cod liver oil, for example, is clearly a food. But if it is intended to treat rickets, it is also a drug. However, the newly added part (3) of the drug definition states, "articles (other than food) intended to affect the structure or any function of the body...." In this case, the article in question must be *either* a food or a drug. If it is a food, then it is not a drug, and vice versa. So, for example, if a product meets the legal definition of a food, and it is sold as a product to help an individual lose weight (treat obesity), then it cannot be considered a drug. On the other hand, if it does not meet the legal definition of a food (for example, if it is in pill form and does not provide any taste or nutritional value), and it is being sold as a product to help an individual lose weight, then it would meet this legal definition of a drug (not a food).

The FDA has in the past relied on this part of the drug definition to take legal action against products that it felt were drugs masquerading as foods. Perhaps, the most famous of these cases occurred in the early 1980s against a product known as "Starch Blockers." Starch Blockers were derived from raw kidney beans and sold in tablet or capsule form. Raw kidney beans contain a protein that inhibits the digestion of some carbohydrates, preventing their absorption into the body and consequently reducing the calories that would normally be derived from a carbohydrate-containing food. Thus, this product was being sold to aid in weight loss. If it met the legal definition of a food (intended to alter the structure or function of the body), then it could not be considered a drug. But if it did not meet the legal definition of a food, then it would be a drug (that alters the structure or function of the body). If it was a drug, then the manufacturer would need to meet the FDA requirements for a new drug and prove both the safety and effectiveness of the Starch Blockers product before it could be released on the market. This, of course, would cost the manufacturer perhaps millions of dollars in scientific and clinical testing. And it may be very likely that this scientific testing would not establish the products' safety and effectiveness, thus preventing its sale altogether. Clearly, it was in the manufacturer's best interest to argue that Starch Blockers met the legal definition of a food. They based their argument on the fact that the product was derived from a food (raw kidney beans). The FDA contended that the product did not meet the legal definition of a food (it did not provide taste, aroma, or nutritive value), and therefore it was a drug. The court judged in favor of the FDA. In the box that follows, I have included some excerpts from the court's Findings of Fact and Conclusions of Law.

The central issue in this case involves a determination of whether starch blockers are a drug under 21 USC § 321(g) or a food under 21 USC § 321(f). If a drug, the manufacturers of starch blockers would be required to file a new drug application pursuant to 21 USC 355 and to be regulated as such. The immediate consequence of such a determination would be the issuance of a permanent injunction requiring plaintiffs to remove the product from the marketplace until approved as a drug by the FDA.

Because of the breadth and necessary vagueness of these statutory definitions, it is incumbent upon this Court to formulate usable working definitions for these terms which can be applied to the case at bar.

The plaintiffs have urged the Court to make its determination of whether starch blockers are foods or drugs based upon the source from which the product has been derived and, apparently, upon the common perception of the category into which the main component of the product falls. Thus, it is argued that, as the product is manufactured from beans, indisputably a natural food, and is made up of mere protein, a substance often regarded as a food, the product must be considered a food. This argument must, however, be rejected.

That a product is naturally occurring or derived from a natural food does not preclude its regulation as a drug. Nor does the fact that an item might, in one instance, be regarded as a food prevent it from being regulated as a drug in another. Therefore, that the product is derived from a natural food and is comprised of vegetable protein does not necessitate a finding by the Court that starch blockers are foods.

Resolution of the issue before the Court must come down to a question of intended use. If a product is intended by the user and the manufacturer or distributor to be used as a drug, it will be regulated as such. Conversely, if it is intended that the product be used as a food, it will be so considered.

By its language, the Act contemplates that "food" refers only to those items actually and solely "used for food." 21 USC § 321(f)(1). Expert testimony received in the instant case leads the Court to conclude that substances used for food are those consumed either for taste, aroma, or nutritional value. It is clear that starch blockers are used for none of these purposes. The user of starch blockers uses them as a drug, not as a food.

Nutrilab, Inc. v. Schweiker 547 F.Supp. 880 (1982)

This was an important legal victory for the FDA. The agency succeeded in removing a product from the market that it felt made unsubstantiated claims by having the courts declare that the product met the legal definition of a drug, and not a food. Interestingly, if you were to do a quick Google search for "Starch Blockers" today, you would find a number of similar products being sold openly *and legally*. These products are essentially the same as the raw kidney bean-derived product sold and subsequently banned in the 1980s. Why are these products not considered illegal drugs today? That will become clear a bit later in this book.

Although the "food" and "drug" definitions were helpful in distinguishing, for legal enforcement purposes, these two categories of products, the original 1938 FD&C Act still did not explicitly define where vitamin supplements fell within the law. Did the law consider them to be a food or a drug? Well, the 1938 law included an additional section that provided the FDA with a justification for how to classify vitamin products. This section can be found in the "Misbranded Foods" provisions of the law, Sec 343(j).

21 USC § 343(j) Representation for special dietary use
A food shall be deemed to be misbranded—
If it purports to be or is represented for special dietary uses, unless its label bears such information concerning its vitamin, mineral, and other dietary properties as the Administrator determines to be, and by regulations prescribes as, necessary in order fully to inform purchasers as to its value for such uses.

When Congress wrote this section of the law, they were probably intending that it be used to regulate actual foods (not supplements) that were formulated and intended for use in individuals with special dietary needs, such as infants or diabetics, for example. It is also important to note that this section of the law only related to the labeling and possible misbranding of these food products, not the safety or adulteration of the products. In 1938, there was no mandatory nutrition labeling requirement, as there is today (as a result of the passage of the Nutrition Labeling and Education Act of 1990). This section of the 1938 law simply required that these types of foods (that is, foods for special dietary use) be required to include nutrition labeling information. But by the early 1940s, vitamin supplements were becoming increasingly popular consumer products, and the FDA made the decision to extend the "foods for special dietary use" designation to these products. In 1941, the agency issued a final rule in the Federal Register[3] essentially establishing this classification. The new final regulation stated the following:

> **21 CFR § 2.10** General.
> (a) The term "special dietary uses", as applied to food for man, means particular (as distinguished from general) uses as food, as follows:
> (3) Uses for supplementing or fortifying the ordinary or usual diet with any vitamin, mineral, or other dietary property.

As a result of this new regulation, vitamin products would now also be required to include nutrition labeling information. But even this new final rule did not explicitly mention the terms "vitamin supplement" or "dietary supplement" as it relates to foods for "special dietary use." That did not occur until more than 30 years later, when the FDA formally modified the definition of "special dietary use" foods to include these supplements.[4]

> **21 CFR § 125.1(a)** Definitions and Interpretation of Terms
> (a) The term "special dietary use" as applied to food (including dietary supplements) used by man means a particular use for which an article purports or is represented to be used, including but not limited to the following:
> (2) Supplying a vitamin, mineral, or other dietary property for use by man to supplement his diet by increasing the total dietary intake..."

Thus, the FDA was effectively regulating vitamin/dietary supplements as foods for "special dietary use" since 1941, but did not officially include them within the regulatory definition of "special dietary use" foods until 1973. But how was the FDA actually regulating these "foods for special dietary use" (including supplements) at this time? In 1941, the agency finalized its "special dietary use" regulations that would be the basis for its enforcement strategy for the next few decades.[3] The 1941 regulations established "minimum daily requirements" (MDRs) as the amounts of specific vitamins or minerals needed to prevent nutrient deficiencies. The labeling of "foods for special dietary use" must indicate what percent of the MDR the product supplies. If the product contained

---

[3] 6 FR 5921 (1941).
[4] 38 FR 2149 (1973).

additional substances for which there was no MDR established, the label must include a statement that *"the need for such substance was not established."*

For approximately the next 20 years, the FDA relied on these regulations to initiate numerous successful misbranding actions against vitamin/mineral products that made disease claims. But beginning in the 1960s, the agency became increasingly concerned about the rise in health and nutrition quackery and the proliferation of supplement products that it felt deceived the consumer with unsubstantiated promises of disease preventing or treating benefits. It also became apparent to the agency that, although they had been successful at taking action against vitamin and mineral supplements that made disease claims on their labeling, this approach was no longer enough to adequately protect the consumer. As Lewis Grossman notes in his review of the history of dietary supplement regulation,[1] by the 1960s, claims made on the labeling of products may have had less impact on consumer purchasing decisions related to dietary supplements, then the "nutrition mythology" culture that existed in America at this time. The public was inundated with nutrition misinformation and quackery via best-selling authors (consider Linus Pauling and his popular book, *Vitamin C and the Common Cold*), television talk shows, and other popular media sources. In addition, there was growing interest in "natural" remedies and growing suspicion of mainstream medicine and "Big Pharma." The FDA decided to embark on a much more aggressive regulatory approach to combat this problem.

In June of 1962, the FDA published in the Federal Register new *proposed* regulations for "special dietary use" foods.[5] The proposed regulations would replace the previous "MDR" with simply the term, "daily requirement." The agency felt that including the word *minimum* may have suggested to some consumers that more was better. The proposed regulations would also permit the label to list *"only those nutrients recognized by competent authorities as essential and of significant dietary-supplement value in human nutrition and that are present in amounts that are consistent with the nutritional requirements for such nutrients."* It then went on to specify the 12 nutrients that it felt met this requirement. Other nutrients or other substances in the supplement would not be allowed to be listed on the label. To most nutritional scientists (then and now) and most other competent health professionals, this seemed like a very logical, reasonable, and scientifically sound approach. However, these proposed regulations were not received well by the dietary supplement industry. The FDA received over 54,000 comments opposing the new regulations, the largest response ever received by the FDA regarding proposed regulations. But as William Goodrich, General Counsel for the FDA at that time, noted, over 40,000 of these responses were

---

[5] 27 FR 5815-5818 (1962).

post cards generated by the National Health Federation (a dietary supplement advocacy and lobbying organization), and approximately 12,000 of the letters were based on a format suggested in Prevention Magazine.[6] Clearly, the dietary supplement industry knew how to mobilize its base.

Goodrich's 1964 review of this topic also included a powerful and strongly worded explanation for the agency's intentions with the new regulations.[7] In the box below is a particularly relevant excerpt from this article.

What was proposed was to eliminate some of the basic causes of confusion and misrepresentation, so that the unwary consumer would be spared the snares which now influence his purchase. What are these sources of confusion?

1. Formulations of vitamin-mineral tablets which contain many times the "minimum daily requirement" of the nutrients, designed to appeal to the layman's belief that if a little is good to satisfy his "minimum" daily requirement, then much more than that minimum would be much better for his health.

2. Formulations based on no rational nutritional principles at all, for example, those which might have one-tenth of the minimum daily requirement of an expensive nutrient and ten times the minimum daily requirement of the cheaper ones.

3. So-called "shotgun" preparations, which contain not only all the vitamins and minerals which might possibly play a part in improving the nutritive well-being of the customer, but also all the other nutrients so far discovered, regardless of the fact that there is no evidence whatever to support a belief that they are needed in human nutrition. And to top this off, some formulations contain a secret base of alfalfa or yeast or some dried animal glands, on the theory that this may contain a possible, as yet undiscovered, vitamin or mineral. The formulators apparently have borrowed a leaf from the book of the ancients, who wanted to appease all the gods by erecting a statue to them and who, when they had erected a statue to all the known gods, then, fearing they may have overlooked one, erected another statue to the unknown god. Nothing is overlooked. Improvement is promised to eliminate every possible nutritional deficiency, established and unestablished, known and unknown.

4. And finally, formulations to support promotional efforts based on the four great myths of nutrition: (a) that our soils are so depleted that ordinary

[6] Goodrich, WW. The Coming Struggle over Vitamin-Mineral Pills. *Bus. Lawyer,* 20(1): 145–150 (1964).
[7] Ibid.

> *foods do not contain the expected nutrients; (b) that modern processing and storage of foods strips them of virtually all-important nutritive values; (c) that it is essentially impossible to obtain from our daily diets the nutrients we require; (d) and that as a result almost everyone is now or will soon be suffering from a subclinical nutritional deficiency which may be the cause of some serious condition of ill health."*
>
> William Goodrich. The Coming Struggle over Vitamin-Mineral Pills.
> Business Law 20(1):145–150 (1964)

The FDA did not get around to publishing its "final" regulations until 1966.[8] And despite the very large response received by the agency in opposition to its 1962 proposed regulations, the FDA actually "doubled down" on the restrictions included in its 1966 final regulations. Among the major provisions in these new final regulations:

1. The MDRs would now be changed to "U.S. Recommended Daily Allowances (USRDA)," and based on the previously established Recommended Dietary Allowances (RDAs) of the Food and Nutrition Board of the National Academy of Sciences.
2. "Standards of Identity" would be established, limiting which nutrients could be included in a supplement and at what levels they could be included.
3. A "Standard of Identity" would be established that would specify the foods to which nutrients could be added.
4. The labeling of these products would not be allowed to make any claim related to the so-called "four great myths of nutrition" that Goodrich identified (see item 4 in the box earlier).
5. And certainly the most controversial provision of the new regulations was the requirement that all dietary supplement labels bear what was known as the "crepe label" statement, specified in the box later.

> *Vitamins and minerals are supplied in abundant amounts by the foods we eat. The Food and Nutrition Board of the National Research Council recommends that dietary needs be satisfied by foods. Except for persons with special medical needs, there is no scientific basis for recommending routine use of dietary supplements.*

---

[8] 31 FR 8521-8527 (1966).

As with the 1962 proposed regulations, these 1966 final regulations seem quite reasonable and scientifically sound from the perspective of nutrition scientists and other creditable health professionals. But the "crepe label" requirement was a very bold step for the FDA and was certain to generate a very strong response from the dietary supplement industry and from the millions of supplement users. By requiring dietary supplement manufacturers to include this statement on the labels of its products, the FDA was essentially forcing the industry to tell their potential-buying public that they really don't need this product. In effect, they are wasting their money if they purchased the product.

Of course, the FDA was inundated with objections to the new regulations and was forced to postpone their enactment while hearings were held to allow more extensive public comment and input. The hearings lasted for 2 years (1968–1970) and resulted in more than 30,000 pages of testimony. Finally, in 1973, based in part on the information obtained from the hearings, the FDA once again attempted to issue updated "special dietary use" food regulations.[9] The agency believed it was offering a more moderate approach that would be more favorably received by the dietary supplement industry (relative to the 1966 regulations). The major provisions of the 1973 regulations were as follows:

1. The FDA dropped the highly controversial "crepe label" requirement.
2. They adopted the USRDA as the basis for how much of the nutrient was supplied in the product.
3. Products were required to be labeled with the percent of the USRDA that the nutrient supplied.
4. Products were still prohibited from making any claims related to the "four great myths of nutrition."
5. The FDA would establish "Standards of Identity" for various types of vitamin and/or mineral supplements (again requiring that dietary supplement products contain only specific vitamins or minerals listed as part of the Standard of Identity).
6. Dietary supplement products must contain a minimum of 50% of the USRDA for the allowed specified vitamin or mineral.
7. Dietary supplement products would not be permitted to contain more than 150% of the USRDA for the specified vitamin or mineral.
8. Any product that contained in excess of 150% of the USRDA would be classified as a drug.

Once again, the FDA's rationale for these updated regulations would likely have seemed reasonable and scientifically sound by nutrition and health

[9] 38 FR 20708-20718 (1973).

professionals. As with the 1962 and 1966 regulations, this latest version was still intended to "eliminate the confusion and misrepresentation" identified in the Goodrich paper.[6] And while the FDA conceded the elimination of the "crepe label" requirement (perhaps hoping that would be enough to appease the opponents of the 1966 regulations), it retained the "Standards of Identity" limiting which nutrients could be added to these products, and it actually went a step further by setting the minimum and maximum levels of nutrients in a supplement product. Setting a minimum level was intended to ensure that these products supplied, at the very least, an amount that would offer some nutritional benefit. Setting the maximum level at 150% of the USRDA was a bit trickier to justify. The FDA argued that there is no scientific basis for a *nutritional* benefit or purpose for any of the allowable nutrients in excess of 150% of the USRDA. And, in fact, if a manufacturer produces a product that contains in excess of 150% of the USRDA, then they must be *implicitly* suggesting that it offers some benefit beyond basic nutrition ... in other words, a disease treating or preventing benefit. This would be true even if the manufacturer did not include an explicit therapeutic claim on the labeling. The fact that it contains such a high level of a nutrient, and that these high levels are not nutritionally necessary, *implies* a therapeutic claim. Furthermore, the FDA argued that some nutrients (such as fat-soluble vitamins A and D) can be toxic at high levels.

Again, there was immediate and massive opposition to the new regulations, including challenges in court. The announcement of these regulations in the Federal Register in 1973 may be viewed as a major turning point in the decades-long attempts by the FDA to more aggressively regulate what it saw as the "nutritional quackery" associated with much of the dietary supplement industry. For the next 20 years, the momentum would shift strongly in favor of the industry, as we shall see in later chapters of this book. This shift began immediately following FDA's 1973 regulation announcement, with important and far-reaching actions occurring at both the legislative and judicial levels. In a landmark case (National Nutritional Foods Association v FDA, US Court of Appeals, 1974), the court found that the FDA had exceeded its authority in both setting standards of identity for dietary supplements (limiting which nutrients could be included in a dietary supplement), and for defining as drugs any supplement that provided more than 150% of the USRDA. The court did uphold the FDA's prohibition against disease claims on dietary supplements and against making any claims related to the "four great myths of nutrition." Nevertheless, this was a significant judicial victory for the supplement industry. Three years later, the FDA again lost in court when it attempted to classify high-dose vitamin A and D supplements as drugs based on their toxicity risk and lack of nutritional value at these doses. With these two judicial decisions, the courts were making it very clear that dietary supplements were food, and that unless the manufacturer made an explicit claim that the product was "intended" to treat, cure, or prevent a disease, the FDA could not regulate them as drugs.

And if these FDA defeats in the courts were not enough, the U.S. Congress also moved to radically limit the authority of the FDA to regulate dietary supplements. In 1976, Senator William Proxmire (D-Wisconsin), a strong advocate for the dietary supplement industry and a personal believer in the health benefits of these products, sponsored an amendment to the FD&C Act. This Vitamin and Mineral Amendment of 1976 (also popularly known as the Proxmire Amendment) is presented in the box below.

21 USC § 350. Vitamins and minerals
  (a) Authority and limitations of Secretary; applicability
    (1) Except as provided in paragraph (2)—
      (A) the Secretary may not establish, under section 321(n), 341, or 343 of this title, maximum limits on the potency of any synthetic or natural vitamin or mineral within a food to which this section applies;
      (B) the Secretary may not classify any natural or synthetic vitamin or mineral (or combination thereof) as a drug solely because it exceeds the level of potency which the Secretary determines is nutritionally rational or useful;
      (C) the Secretary may not limit, under section 321(n), 341, or 343 of this title, the combination or number of any synthetic or natural—
        (i) vitamin,
        (ii) mineral, or
        (iii) other ingredient of food, within a food to which this section applies.

In the earlier box, the term "Secretary" is referring to the head of the Department of Health and Human Services (in 1976 known as the Department of Health, Education, and Welfare). As was discussed earlier, the FDA is part of HHS, so this amendment is a direct Congressional rebuke of the FDA and its efforts to regulate dietary supplements. The phrasing of this amendment is remarkably strong, with a tone that makes it clear that the Congress was not pleased with the regulatory approaches that the FDA had unsuccessfully attempted over the previous 20 years. This section remains part of the FD&C Act today, although as we will see a bit later in this book, it has been expanded somewhat as the result of the passage of the 1994 Dietary Supplement Health and Education Act.

Following these defeats in both the courts and the Congress, the FDA finally officially revoked its "Foods for Special Dietary Use/Vitamin and Mineral Products" regulations on March 6, 1979.[10] However, this was certainly not the end of the long ongoing battle between the FDA and the dietary supplement industry. Over the next 15 years, the FDA would struggle with new safety, policy, advertising/labeling, and societal implications related to dietary supplement regulation. Coming up in this book, we will explore how a relatively unexciting advertisement for a breakfast cereal had an enormous impact on how dietary supplements are manufactured, labeled, and marketed today. But first we need to briefly review the role of another executive branch agency that played (and continues to play) an important role in the dietary supplement story: the Federal Trade Commission.

---

[10] 44 FR 16005-16006 (1979).

chapter six

Food labeling versus food advertising

As a result of the court decisions and legislative actions of the mid to late 1970s, the Food and Drug Administration's (FDA's) authority to regulate dietary supplements was greatly restricted. It is important to remember, however, that neither the court decisions nor the Vitamin and Mineral Amendment of 1976 prevented the FDA from declaring any dietary supplement to be a drug if it made a claim on its labeling that it was intended to treat, cure, or prevent any disease. But what about the *advertising* of a dietary supplement product? The FDA can regulate product labeling but has no authority to regulate advertising. That authority lies with a different executive branch agency; the Federal Trade Commission or FTC. The FTC's authority to regulate product advertising results from the passage of the FTC Act in 1914. Among the major provisions of this law is a prohibition against "false advertisements" that are "unfair or deceptive." A bit later in this section, we will review the criteria for determining when an ad is considered *unfair or deceptive*. But first we need to distinguish between what is an advertisement and what is labeling.

The FTC Act does not explicitly define the term *advertisement*. A Google search for a "legal" definition of advertisement will yield a variety of results, but they all have a similar general theme. That is, an *advertisement* is any form of communication (print, radio, television (TV), display, website, etc.) intended to influence consumer attitudes regarding the purchase of a particular product. But what about the printed material or images on the package of a food product itself? Is that not also a form of communication intended to influence the purchase of the product? Yes, but from a legal standpoint, it is defined and regulated differently.

The Food, Drug, and Cosmetic (FD&C) Act of 1938 defined two important terms: food *label* and food *labeling*.

21 U.S. code (USC) § 321: *Definitions*

(k) The term "label" means a display of written, printed, or graphic matter upon the immediate container of any article; ...

(m) The term "labeling" means all labels and other written, printed, or graphic matter (1) upon any article or any of its containers or wrappers, or (2) accompanying such article.

This legal definition distinction between *label* and *labeling* may seem a bit trivial or confusing at first glance, but from a food regulation perspective it is very important. The first important point to remember is that the FDA—not the FTC—regulates both food labels and food labeling. The second important point to note is that the legal definition of a food "label" is essentially a common sense definition (*"a display of written, printed, or graphic matter upon the immediate container of any article"*). This is probably what most of us think of as a food label. But the *labeling* definition is a bit trickier. As the definition clearly states, all *labels* are part of the *labeling* of a product. However, the labeling of a product also includes any *"... written, printed, or graphic matter... accompanying such article."* This means that any material "near" the product, or *in proximity* to the product, and that is clearly associated in some way with the product, is considered part of the labeling of the product and can be regulated by the FDA. For example, consider a hypothetical dietary supplement product that makes no claim on its *label* that it is intended to treat, cure, or prevent a disease. However, in the store where the supplement is sold, it is displayed alongside a brochure or article that talks about a relationship between the active ingredient in the supplement and its effects on a particular disease. This brochure or article would legally be considered part of the *labeling* of the product. As a result, the FDA could determine that the supplement met the legal definition of a drug by virtue of the disease claim in the labeling of the product. (At least, that was the case before the passage of the Dietary Supplement Health and Education Act (DSHEA) in 1994, as we shall see a bit later in this book).

Sometimes, however, the FDA's regulation of labeling and the FTC's regulation of advertising can overlap. In the dietary supplement example earlier, the FTC could also consider the brochure to be an advertisement for the product. Or perhaps there is a display near the product that shows a picture of a celebrity touting the benefits of the product. That display would certainly meet the definition of labeling (since it is *accompanying* the product) and could also be considered an advertisement. This could result in a conflict as to which agency should take an action against the product if the FDA feels that the display represents labeling that is *misbranded* according to the FD&C Act, and the FTC feels that the same display represents an *unfair or deceptive* advertisement according to the FTC Act. The U.S. Congress anticipated this potential problem when they wrote the FD&C Act in 1938 and included a provision intended to address this potential conflict. This provision can be found in 21 USC § 378. I have paraphrased this section of the law in the box below.

> **Paraphrased summary of 21 USC § 378: Procedures to be taken when FDA believes labeling is misbranded, but the labeling could also be considered "advertising" by the FTC.**
>
> 1. The FDA would notify the FTC that the advertisement constitutes misbranding of food labeling, and what action the FDA intends to take against the product, along with supporting documents and evidence.
>
> 2. Within 30 days, the FTC must notify the FDA as to whether or not it has initiated an investigation into the product advertising, or has begun to take some action against the company, or it has issued a complaint against the company. If the FTC has done any of these things, then the FDA may not act against the product or company for misbranding.
>
> 3. If, after 60 days, the FTC has not done any of the earlier steps, then the FDA may initiate its own action with respect to the misbranding of the food labeling.
>
> 4. If the FDA feels that the misbranding of the food labeling presents an imminent hazard to health, then it does not need to go through all of the earlier steps. Rather, it can act immediately to eliminate the hazard.

This coordination between the FDA's regulation of labeling and the FTC's regulation of advertising has proven at times to be quite a powerful enforcement collaboration. For example, we have just finished reviewing how the court decisions and legislative actions of the 1970s greatly limited the ability of the FDA to regulate dietary supplements within the constraints of the "misbranding" provisions of the FD&C Act. However, the FTC Act gives the FTC a much broader latitude in determining when an advertisement is unfair or deceptive. Therefore, if the FDA is unable to act against a product based on the misbranding of its labeling, the FTC may be able to act against the same product based on the unfair or deceptive criteria of the product's advertisement. The next obvious question is what constitutes an unfair or deceptive advertisement. Let us start by looking at some pertinent sections of the FTC Act.

15 USC §45(a)(1)
Unfair methods of competition unlawful; prevention by Commission
(1) Unfair methods of competition in or affecting commerce, and unfair or deceptive acts or practices in or affecting commerce, are hereby declared unlawful.

Notice that the above portion of the FTC Act is found in Title 15 of the USC (not Title 21 where you would find the FD&C Act). This section of the law is fairly straightforward and clear. It simply declares that unfair or deceptive acts or practices related to how a company sells its products are unlawful. Advertising is certainly an "act or practice... affecting commerce".

The next section goes on to clearly state that it is illegal to disseminate any advertisement that is "false."

15 USC §52(a)(2)
Dissemination of false advertisements.
It shall be unlawful for any person, partnership, or corporation to disseminate, or cause to be disseminated, any false advertisement—
(2) By any means, for the purpose of inducing, or which is likely to induce, directly or indirectly, the purchase in or having an effect upon commerce, of food, drugs, devices, services, or cosmetics.

The next section simply makes it clear that a "false advertisement" is, by definition, an unfair or deceptive act or practice.

15 USC §52(a)(2)
Dissemination of false advertisements
(b) Unfair or deceptive act or practice
The dissemination or the causing to be disseminated of any false advertisement within the provisions of subsection (a) of this section shall be an unfair or deceptive act or practice in or affecting commerce within the meaning of section 45 of this title.

And finally, the next section describes some of the criteria that may be considered when determining whether an advertisement is "false" and consequently "unfair or deceptive."

> **15 USC §55(b).** Additional definitions
> For the purposes of sections 52–54 of this title—
> (a) False advertisement
> (1) The term "false advertisement" means an advertisement, other than labeling, which is misleading in a material respect; and in determining whether any advertisement is misleading, there shall be taken into account (among other things) not only representations made or suggested by statement, word, design, device, sound, or any combination thereof but also the extent to which the advertisement fails to reveal facts material in the light of such representations or material with respect to consequences, which may result from the use of the commodity to which the advertisement relates under the conditions prescribed in said advertisement, or under such conditions as are customary or usual.

This last quoted section of the law is particularly interesting and relevant. It hints to the broad latitude that I mentioned earlier regarding the FTC's authority. In determining whether an advertisement is *unfair or deceptive*, the FTC can consider representations "made or suggested." And the FTC can consider not only what information is included in the ad but also what information is not included ("*...the extent to which the advertisement fails to reveal facts...*"). The FTC's website provides additional details to help companies avoid unfair or deceptive advertising. This information is in the Frequently Asked Questions section of the FTC website[1] and some of this information is reproduced in the box that follows.

> **What makes an advertisement deceptive?**
>
> According to the FTC's Deception Policy Statement, an ad is deceptive if it contains a statement—or omits information—that
>
> Is likely to mislead consumers acting reasonably under the circumstances and
>
> Is "material"—that is, important to a consumer's decision to buy or use the product.

---

[1] Federal Trade Commission website, "Advertising FAQ's: A Guide for Small Business." www. ftc.gov/tips-advice/business-center/guidance/advertising-faqs-guide-small-business.

What makes an advertisement unfair?

According to the FTC Act and the FTC's Unfairness Policy Statement, an ad or business practice is unfair if

It causes or is likely to cause substantial consumer injury which a consumer could not reasonably avoid and

It is not outweighed by the benefit to consumers.

How does the FTC determine whether an ad is deceptive?

A typical inquiry follows these steps:

The FTC looks at the ad from the point of view of the "reasonable consumer"—the typical person looking at the ad. Rather than focusing on certain words, the FTC looks at the ad in context— words, phrases, and pictures—to determine what it conveys to consumers.

The FTC looks at both "express" and "implied" claims. An express claim is literally made in the ad. For example, "ABC Mouthwash prevents colds" is an express claim that the product will prevent colds. An implied claim is one made indirectly or by inference. "ABC Mouthwash kills the germs that cause colds" contains an implied claim that the product will prevent colds. Although the ad doesn't literally say that the product prevents colds, it would be reasonable for a consumer to conclude from the statement "kills the germs that cause colds" that the product will prevent colds. Under the law, advertisers must have proof to backup express and implied claims that consumers take from an ad.

The FTC looks at what the ad does not say—that is, if the failure to include information leaves consumers with a misimpression about the product. For example, if a company advertised a collection of books, the ad would be deceptive if it did not disclose that consumers actually would receive abridged versions of the books.

The FTC looks at whether the claim would be "material"—that is, important to a consumer's decision to buy or use the product. Examples of material claims are representations about a product's performance, features, safety, price, or effectiveness.

> The FTC looks at whether the advertiser has sufficient evidence to support the claims in the ad. The law requires that advertisers have proof before the ad runs.
>
> From: Federal Trade Commission website, "Advertising FAQ's: A Guide for Small Business."
> www.ftc.gov/tips-advice/business-center/guidance/
> advertising-faqs-guide-small-business.

Thus, in deciding whether an ad is unfair or deceptive, the FTC will consider the ad in its entirety. What is the net impression of the ad to the general public, or to the particular target audience (such as children), or even to the more gullible, ignorant, or unthinking members of the target audience? The FTC does not need to prove that there was intent to deceive on the part of the company. It just needs to prove the capacity of the ad to deceive. Inconspicuous disclaimers or technical jargon will not save a deceptive ad from FTC action.

In addition, the FTC also has some powerful enforcement options that are not available to the FDA. Its principle enforcement tool is a "cease and desist order." This is simply an order handed down from the FTC's Administrative Law Judge (after being presented with all testimony and evidence), ordering the company to stop engaging in a particular activity. The FTC can also require the company to produce "corrective advertising" if it feels that is the only way to *erase* any lingering negative effects of the false advertisement among consumers. And the FTC can order the company to pay restitution to individuals monetarily harmed as a result of the false advertisement.

All of this may appear to suggest that the FTC has all the tools it need to prevent any unfair or deceptive advertisements from reaching the public, whether it be from dietary supplements or any other consumer product. However, while it may have the enforcement tools, it does not have a budget nor the personnel resources to effectively monitor and prosecute the enormous number of potentially false advertisements produced each year in the United States. It is forced, therefore, to focus its efforts on the most egregious offenders and those that are inflicting the most harm to consumers.

One recent FTC action is particularly relevant to this book, as it involves a product that I mentioned in the introduction chapter. Prevagen is a dietary supplement advertised to improve memory and cognition. The makers of Prevagen (Quincy Bioscience) have advertised heavily on TV commercials, infomercials, the web, and elsewhere.

As you can see from the packaging image, it does not make any disease claim. It does make some "structure/function" claims (Healthy Brain Function, Sharper Mind, Clearer Thinking), but these are permitted under the DSHEA of 1994 (as we shall see later in this book). As far as the FDA is concerned, this product does not violate the FD&C Act. Yet in January of 2017, the FTC, along with the New York State Attorney General, charged the makers of Prevagen with making false and unsubstantiated claims about the product.

The advertising for Prevagen quotes the results of a "randomized, double-blind, placebo-controlled clinical trial" to prove that the product does indeed improve memory and cognition. For scientists, these types of studies are, in fact, considered to be the "gold standard" for determining whether a substance is effective in accomplishing what it is intended to do—in this case, to improve memory. But the FTC and the New York State Attorney General contend that the study is fraught with design problems, that the results of the study do not prove that Prevagen works (and instead may, in fact, prove that it doesn't work), and that the company misrepresented the results of the study in its advertisements for the product. In the box that follows, I have quoted some of the excerpts from the FTC complaint, detailing why the FTC considers Prevagen's ads to be unfair and deceptive.

> *To substantiate their claims that Prevagen improves memory, is clinically shown to improve memory, improves memory within 90 days, is clinically shown to improve memory within 90 days, reduces memory problems associated with aging, is clinically shown to reduce memory problems associated with aging, provides other cognitive benefits, and is clinically shown to provide other cognitive benefits, Defendants primarily rely on one double-blind, placebo-controlled human clinical study using objective*

outcome measures of cognitive function. This study, called the Madison Memory Study, involved 218 subjects taking either 10 milligrams of Prevagen or a placebo. The subjects were assessed on nine computerized cognitive tasks, designed to assess a variety of cognitive skills, including memory and learning, at various intervals over a period of ninety days. The Madison Memory Study failed to show a statistically significant improvement in the treatment group over the placebo group on any of the nine computerized cognitive tasks.

After failing to find a treatment effect for the sample as a whole, the researchers conducted more than 30 post hoc analyses of the results, looking at data broken down by several variations of smaller subgroups for each of the nine computerized cognitive tasks. This methodology greatly increases the probability that some statistically significant differences would occur by chance alone. Even so, the vast majority of these post hoc comparisons failed to show statistical significance between the treatment and placebo groups. Given the sheer number of comparisons run and the fact that they were post hoc, the few positive findings on isolated tasks for small subgroups of the study population do not provide reliable evidence of a treatment effect.

Nevertheless, Defendants widely touted the Madison Memory Study in their advertising. For example, the chart below appeared in the product labels for the Prevagen Products and Defendants' TV ads and website, prevagen.com. It indicates that a "double-blinded, placebo controlled study" showed dramatic improvement in recall tasks when, in fact, the results for the specific task referenced in the chart showed no statistically significant improvement in subjects taking Prevagen compared to subjects taking a placebo. In addition, Defendants eliminated from the chart one of the four data points in the study – day 60. At day 60, the recall task scores of subjects taking Prevagen declined from day 30, and were slightly worse than the recall task scores of subjects in the placebo group.

Defendants' claims that their product improves memory and cognition rely on the theory that the product's dietary protein, apoaequorin, enters the human brain to supplement endogenous proteins that are lost during the natural process of aging. Defendants developed their product and created their marketing campaign based

> *on this theory. Defendants, however, do not have studies showing that orally-administered apoaequorin can cross the human blood brain barrier and therefore do not have evidence that apoaequorin enters the human brain. To the contrary, Defendants' safety studies show that apoaequorin is rapidly digested in the stomach and broken down into amino acids and small peptides like any other dietary protein.*
>
> FTC and New York State Attorney General v. Quincy Bioscience Holding Company.
> Complaint for Permanent Injunction and Other Equitable Relief.
> www.ftc.gov/system/files/documents/cases/quincy_bioscience_complaint-filed_version.pdf

The complaint went on to specify the charges against the company, concluding that the advertisements *"... are false or misleading, or were not substantiated at the time the representations were made."* And further, that the making of the representations in the ads *"... constitutes a deceptive act or practice and the making of false advertisements, in or affecting commerce, in violation of sections 5(a) of the FTC Act, 15 USC Sections 45(a) and 52."*

The FTC was seeking to obtain a permanent injunction against the company, along with *"restitution, the refund of monies paid, and disgorgement of ill-gotten monies."* Notice also that in this case, the FTC was not employing a "cease and desist order" (issued by an FTC Administrative Law Judge). Rather it was seeking a permanent injunction (ordering the company to stop the ads). To obtain permanent injunction, the FTC and New York State Attorney General filed a lawsuit against the company in the U.S. District Court. This would involve a trial in which both sides can present their case.

This Prevagen case nicely illustrates the role that the FTC can play in the regulation of dietary supplements. In this case, the FDA really had no legal basis for going after the company, as the company did not violate any of the FD&C Act's labeling/misbranding laws or other regulations that the FDA enforces. However, the FTC was able to file a lawsuit against the company, charging the company for violation of the FTC Act's provisions against false advertisements that are, by definition, unfair or deceptive. If the FTC is ultimately successful in court, the company may be required to immediately stop disseminating the advertisements, refund money to consumers who purchased the product, and pay back (disgorgement) any profits the company earned as a result of the false advertisements. The FTC's regulatory authority over the advertising of dietary supplements provides an important additional enforcement oversite of this industry. And the FTC website (www.ftc.gov) is an excellent resource for consumer information on dietary supplements and the claims that are made in their advertising. The FTC also provides helpful guidance for industry on how to comply with dietary supplement advertising laws and regulations.[2]

Recently, the FTC had a serious setback in its case against the manufacturers of Prevagen, Quincy Bioscience. The company moved to dismiss the FTC's complaint and in September of 2017, the District Court granted the motion. The arguments on both sides of this case nicely illustrate the challenges that federal regulators and courts face when dealing with the interpretation of scientific data in a legal setting. As was pointed out in the box earlier, the FTC's argument focused on the fact that the original clinical trial with Prevagen failed to show any beneficial effect on nine parameters of cognitive function and memory. But rather than accept these negative results as evidence of the ineffectiveness of their supplement, the Quincy Bioscience-sponsored investigators decided to perform what is known as a *post hoc* analysis. A *post hoc* analysis involves reanalyzing the study with additional new research comparisons; comparisons that were not intended to be asked within the original experimental design. When they now reanalyzed the data from these more than 30 new comparisons, they found statistically significant results for three of the measurements (27 results were still not significant). But this type of *post hoc* "multiple comparison" statistical analyses is generally considered to be very bad science. This is because in any scientific experiment, there is always the possibility (probability) that you will get a positive result simply by chance. In other words, you can never be 100% sure that the result you are observing is due to the effect of the drug or supplement. Statistical analyses allow researchers to assign a probability to the likelihood that the result they observed was due to chance, and not due to the intended effect of the treatment (in our case, the intended effect of the Prevagen supplement). In the Prevagen study, their original analysis found no treatment benefit from the supplement. But when they later performed many more *post hoc* analyses, the probability that they would get a positive result went up dramatically, even though the likelihood that the positive result was due to chance (and not due to the supplement) also went up dramatically.

This statistical concept can be rather easily explained with a standard dice role analogy. If you were asked what is the probability that you could role a "two" with one role of a single die, you would quickly recognize

---

[2] Dietary Supplements: An Advertising Guide for Industry. U.S. Federal Trade Commission. www.ftc.gov/tips-advice/business-center/guidance/dietary-supplements-advertising-guide-industry.

that this probability is 1 in 6 or 16.7% (with six sides to the die). But if you continue to role the die many times the probability that one of the roles will result in a "two" will go up. For example, if you role the die ten times, the probability that you will role a "two" goes up to 84%. This is effectively what Quincy Bioscience did. They kept measuring comparisons (like continuing to roll the die) until they found the result they wanted. But the likelihood that the result was due to Prevagen (and not just random chance) becomes very small. Unfortunately, there is still always a small probability that the result is, in fact, due to Prevagen. It is this point that the District Court noted in its decision to dismiss the FTC's complaint against Quincy Bioscience. The court determined that the FTC's complaint relied on the theoretical (and statistical) likelihood that the results that Quincy Bioscience used in their advertisements were due to chance (resulting from the multiple comparisons analyses) and not due to a beneficial effect of the Prevagen supplement. In its decision to dismiss the FTC's complaint, the District Court concluded, therefore, that it could not ignore the possibility that the result was true (a beneficial effect of Prevagen), even if the likelihood of this being the case was small (due to the multiple comparison effect).

This is not the only problem with the Quincy Bioscience study and its claim that Prevagen can affect brain cells. The active ingredient in Prevagen is a protein. When you ingest proteins (as you would when you take the Prevagen supplement), the protein in the supplement is broken down into its component "amino acids" by acids and enzymes in the stomach and small intestine. In other words, any possible biological activity of the Prevagen protein would be destroyed during digestion. And there is no scientific evidence to suggest that the protein in Prevagen can escape this effect during digestion. Thus, there is likely no Prevagen-derived protein entering the blood stream to reach the brain. In addition, even if any of the Prevagen protein somehow enters the circulatory system, there is a barrier (known as the blood–brain barrier) that functions to prevent circulating proteins from reaching the brain. Thus, from a nutritional, physiological, and biochemical perspective, it makes little sense that Prevagen can have any effect on brain cells. At the time of writing this book, the FTC is appealing the District Court's dismissal of the original complaint. It is a story that will certainly be worth following.

chapter seven

The battle over health claims

As the 1970s drew to a close, the Food and Drug Administration (FDA) was left with essentially only one enforcement option for dealing with dietary supplements that the agency felt crossed a line into "nutritional quackery." That is, if the dietary supplement made an explicit disease claim, the FDA could determine that it met the legal definition of a drug and, as such, would be deemed misbranded unless it satisfied the premarket proof of safety and effectiveness required of new drugs. During the 1970s and early 1980s, this strategy proved reasonably effective at dealing with the worst offenders; manufacturers that made the most dubious and sometimes outrageous disease preventing, curing, or treating claims about their products. When the FDA might have a weak case based on the "misbranding" provisions of the Food, Drug, and Cosmetic (FD&C) Act, the Federal Trade Commission (FTC) could possibly be effective at going after these products based on the "unfair or deceptive" standards for advertising. Furthermore, the "Starch Blockers" case gave the FDA additional legal precedence for going after products that may not have made explicit disease claims, but were considered drugs if they were substances "other than food" intended to alter the structure of function of the body.

During this same approximate period, there was an expanding interest and discussion among health professionals regarding the need to better communicate the importance of a healthy diet at reducing the risk of various health issues and disease. For example, was there a better way to communicate to the public the relationship between a high-fat diet and the risk of cardiovascular disease? Or was there a better way to communicate to the public the relationship between sodium intake and the risk for hypertension?

Food labeling was prohibited from making any direct or explicit health claims, even for diet/disease relationships that were known at the time to be based on scientifically sound evidence and of public health significance, such as the fat and heart disease relationship, or the sodium and hypertension relationship, and others. In many ways, this rather rigid outright prohibition against *any* disease claim on food product labeling worked to the FDA's advantage. While the FDA would likely have agreed with the importance of communicating, for example, the relationship between sodium intake and hypertension, it also recognized that if it allowed a sodium/hypertension health claim on food labeling, it would be forced to make future determinations about which other health claims are important and scientifically sound. By sticking with the outright ban on all health/disease claims (even those that are important and scientifically sound), it avoided having to make decisions on what would certainly have been an avalanche of more questionable health/disease claims for other products to follow. In addition, the FDA knew that it had the law on its side. The FD&C Act was very clear regarding the labeling of a product making a disease claim. These products meet the legal definition of a drug. And the nutrition labeling regulations reinforced this point, as indicated in the CFR excerpt below.

21 CFR Sec 101.9

(i) A food labeled under the provisions of this section shall be deemed to be misbranded under sections 201(n) and 403(a) of the act if its labeling represents, suggests, or implies:

(1) That the food because of the presence or absence of certain dietary properties, is adequate or effective in the prevention, cure, mitigation, or treatment of any disease or symptom.

However, as we entered the early 1980s, a number of important cultural, political, scientific, and business factors were converging in a way that would ultimately force the FDA away from this rigid interpretation of the drug definition. Culturally, there was an ever-increasing interest in natural remedies and alternative medicine. As I mentioned earlier in this book, this movement started in earnest with the counterculture and antiestablishment attitudes of the 1960s, and continued to expand in the 1970s with more mainstream involvement of celebrities and some fringe controversial scientists (such as Linus Pauling and vitamin C). Politically, we were entering the Reagan era of executive branch deregulation and a movement toward less government interference in the free market system. Scientifically, there was a growing consensus among reputable scientists and health advocacy organizations regarding the roles that various dietary factors play in maintaining health and reducing (or increasing) the risk of many diseases. Finally, the business community was beginning to recognize that there was a significant profit to be made by tapping into consumer interest in health and diet. This, combined with the more favorable attitude toward business of the Reagan political environment and the sound scientific rationale supporting some of the relationships between diet and health, suggested an opportunity for new labeling and advertising strategies.

The FDA was beginning to feel pressure to rethink its rigid position on food product health and disease claims. In practice, the agency had already set some precedent in this regard. It allowed, for example, sugar-free chewing gum to declare that it can help prevent dental cavities. And it allowed some limited health claims for "special dietary use" foods for diabetics or individuals with specific food allergies. In Peter Barton Hutt's 1986 review of the history of health claims in food labeling,[1] he quotes a 1984 talk by the then Acting Commissioner of the FDA, Mark Novitch.

Our problem, as I'm sure you understand, is how to permit appropriate health claims without opening the door to outright fraudulent ones. Current FDA rules prohibit any labeling that makes health claims. But we have allowed labeling such as "no preservatives," "no artificial flavoring," "no artificial coloring," "no caffeine" – which connote that people ought to avoid these things. We have not moved against those kinds of claims because they are truthful and they do not make outright health claims. Similarly, exceptions like permission for sugarless chewing gums to make claims that they don't cause tooth decay have been permitted because they are clearly in the public interest.

So what I am saying today is that we are open to generic health claims – ones about the need for a balanced diet, for moderate caloric intakes, and for avoiding emphasis on foods that make miraculous claims. As for specific claims about specific foods, we need to be cautious and deliberate.

Mark Norvitch, Acting Commissioner of the FDA. "Government's Role in Assuring Truthful Information on Diet and Health." Proceeding of National Food Processors Association Conference on Food Labeling and Advertising. (1984)

---

[1] Hutt, PB. Government Regulation of Health Claims in Food Labeling and Advertising. *Food Drug Cosmet. Law J.*, 41: 3–73 (1986).

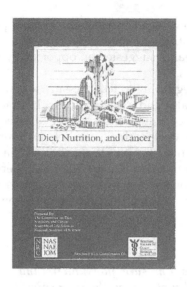

Through speeches such as this and other communications and pronouncements, the FDA was sending clear signals that it may soon allow some carefully controlled use of health claims on food labeling, without triggering the drug definition. Food manufacturers were eagerly anticipating new FDA regulations or guidelines that would permit them to use health claims to promote their products. But despite the FDA's persistent hints that new regulations or guidelines were imminent, by 1984, there was still no official FDA action. There was, however, important movement on the scientific front. In 1980, the National Cancer Institute (NCI) of the National Institutes of Health (NIH) commissioned the National Research Council (part of the National Academy of Sciences) to "conduct a comprehensive study of the scientific information pertaining to the relationship of diet and nutrition to cancer." In 1982, the NRC published its report in a 493-page book titled, *Diet, Nutrition, and Cancer*. This report systematically reviewed the existing epidemiological and experimental data on the relationships of cancer to each of the macronutrients (protein, carbohydrates, fat, fiber), total calories, many of the vitamins and minerals, alcohol, as well as many nonnutritive food components such as food additives, contaminants, and naturally occurring and process-induced toxins in foods. The report also highlighted the strengths and weaknesses of the available scientific data and made recommendations for what additional research was needed.

Then in early 1984, the NCI published another comprehensive review of the diet and cancer relationship (NCI, Cancer Prevention, NIH Pub. No. 84-2671), which included some specific dietary recommendations for reducing your risk of cancer.

Both of these publications attracted widespread interest among scientists, health professionals and educators, the press and popular media, and of course, the food industry. Food manufacturers and advertisers now had government-sponsored studies and reports that included dietary recommendations to guide consumers toward reducing their risk of cancer. But according to current FDA regulations at the time, they were still not allowed to make reference to this information in their labeling and advertising. If they did, the FDA would consider it a "disease claim" and the product could be legally classified as a drug. So, the food industry continued to wait for the FDA to release anticipated new rules or guidelines that would allow food companies to make some explicit health claims on their food labels.

One food company decided it could wait no longer. In the fall of 1984, Kellogg's Company drafted an advertising campaign for its All-Bran breakfast cereal that drew information directly from the NCI's report and dietary recommendations. It worked closely with the NCI in developing the ad campaign and incorporated several of NCI's comments and suggestions into the final ad. For example, NCI wanted to be sure that the ad campaign did not imply or suggest that it was endorsing or directly promoting the product itself. Rather, it was simply endorsing the use of the health message.

There are at least two remarkable and bold aspects to the Kellogg's strategy. First, they did not consult with the FDA during the development of the ad campaign. I think it is fair to assume that Kellogg's knew that the FDA would not have approved of the ad or the associated product labeling and likely would have informed Kellogg's that it would be in violation of the FD&C Act if it went ahead with the ad and labeling. Second, by consulting with the NCI, Kellogg's effectively did an "end run" around the FDA, obtaining an endorsement of the ad (but not the product itself) from a very reputable science organization (the NCI) that is part of the same Executive Branch cabinet-level department as the FDA (both the FDA and the NCI were, and still are part of the Department of Health and Human Services). In addition, Kellogg's likely recognized that the ad campaign would not be opposed by the FTC, as that agency was already on record as supporting limited and scientifically sound health claims in food product advertising. In October of 1984, the All-Bran television (TV) commercials were aired for the first time. An example of one these early commercials is presented in the box on the following page.

These early All-Bran commercials all followed a similar script and style, with a serious actor talking directly into the camera and describing how he/she learned from the NCI about the importance of fiber in the diet. These were certainly not your stereotypical breakfast cereal commercials of the time. In fact, they almost took on more of the feel of Public Service Announcements.

<u>Actor:</u>

I don't usually do this.
But I saw this number on TV and I called it.
It's to the National Cancer Institute.
And what I found out is pretty convincing.
They believe that a high-fiber, low-fat diet may
reduce the risk of some kinds of cancer.
So I made some changes.

Like eating All-Bran.
The natural high-fiber cereal from Kellogg's.
You know … I'm glad I called that number.

<u>Announcer Voiceover:</u>

Kellogg's All-Bran.
For more facts, call the National Cancer Institute,
toll-free.

This was a smart early approach by Kellogg's. The company knew that it was challenging the FDA's long-standing prohibition against health claims for foods and that, technically, the company was breaking the law. In a *New York Times* article at the time, Celeste Clark, director of corporate publicity and nutrition at Kellogg's, defended the ads, telling the newspaper, *"We are saying the message is not a health claim. The message is from the National Cancer Institute, and we are simply printing it."*[2] In fact, that is exactly what Kellogg's did on the back of the All-Bran cereal box, even including contact information where consumers could obtain more information on diet and cancer directly from the NCI (see cereal box labeling on previous page).

Once the TV commercial and the package labeling were released, the FDA was in a very difficult position. A sister agency within the Department of Health and Human Services, the NCI, had not only endorsed the ad campaign, but worked with Kellogg's in its development. The FTC, responsible for regulating advertising and which often worked closely with the FDA to prosecute products that made unsubstantiated health claims, was on record as supporting the Kellogg's advertisement. The Chairman of the FTC at the time, James C. Miller, was very public regarding his disagreement with the FDA on how to deal with the Kellogg's issue. He indicated that the FTC would take a much more lenient approach than the FDA regarding health claims in food advertising, saying, "I think it is very important to have truthful information disseminated."[3] Even the Assistant Health and Human Services Secretary at the time, Edward Brandt, was supportive of the Kellogg's campaign. In his report to the then Secretary of Health and Human Services (HHS), Brandt stated the following:

> *The back panel of the cereal boxes – to be devoted entirely to NCI's messages – will include more detail on the diet recommendations. NCI scientific staff and National Institutes of Health legal advisors reviewed the material to check their accuracy and to be sure that there was no implication that NCI was endorsing the company or its products. This effort by the Kellogg Company will infuse millions of dollars of private sector funds into the promotion of cancer prevention messages.*

Food Chemical News, November 5, 1984

---

[2] New York Times, October 7, 1984. "Health Claims on Food Put FDA in a Corner".
[3] FTC, FDA Don't See Eye-to-Eye on Kellogg Fiber-Cancer Ads, Miller Says. Food Chemical News. December 3, 1984.

So while Kellogg's was clearly violating the law with this labeling, the FDA recognized immediately that there would be no support within HHS or the FTC for taking legal action against the company. Nevertheless, the FDA expressed serious concerns regarding the ads and the labeling. Among its concerns were first, that allowing Kellogg's to continue with this ad campaign without legal consequence, would open a "Pandora's box" of other food or dietary supplement companies making health claims, many of which would likely be completely unsubstantiated or based on dubious science. Second, the FDA believed that while the NCI information in the ads and labeling was technically accurate, it did not convey the complete story and therefore was somewhat misleading. For example, Dr. Sanford Miller, at the time director of the FDA's Center for Food Safety and Applied Nutrition, noted the following regarding the Kellogg's ad campaign:

...one can make a very substantial argument...that there's no evidence that wheat bran or oat bran has anything to do with any of the data collected thus far on colonic cancer, and that all of the data collected thus far on reduction in colonic cancer...is on forms of fiber other than the ones being advertised.

Food Chemical News. February 18, 1985

Dr Miller goes on to point out that the same All-Bran cereal box label includes a recipe for bran muffins that uses an "inordinate amount of butter," and that while the ad mentions the importance of All-Bran as part of a high-fiber, low-fat diet, the nutrition content panel on the box uses whole milk with the cereal. In other words, the Kellogg's All-Bran cereal box labeling was sending consumers a mixed message. The bottom line was that the FDA recognized that while the Kellogg's ads and labeling were not *untruthful*, they are not telling the whole truth. And the agency was concerned that this was going to be a problem with many food product health claims to follow. Thus, while it may be laudable to try to incorporate important health information in food advertising and labeling, it was going to be very difficult to do this accurately and completely in a short TV commercial or on a relatively small food label.

Kellogg's was certainly taking a legal gamble when it decided to challenge the FDA and go ahead with its ad campaign. But the gamble apparently paid off. Not only did the FDA not pursue an enforcement action against Kellogg's, but the ad campaign proved quite profitable for the company. An analysis of the effectiveness of the ads (interestingly, conducted by consumer science specialists at the FDA) found that there were significant increases in consumer purchases of Kellogg's high-fiber cereals, particularly All-Bran, linked to the advertisements.[4] According to this study, Kellogg's spent more than $28 million on high-fiber cereal advertising in 1985, and 50% of this was for its All-Bran cereal. Kellogg's claimed that as a result of their ad campaign, over 90% of adults knew of the fiber-cancer connection and heard the message an average of 35 times each.[5] It is certainly hard to argue that the ad campaign was not effective at getting the NCI message out to the public. And the food industry now had hard data to demonstrate that health claims on food labels and in food advertising provides consumers with important health information and that the partnership between the food industry and health promotion organizations can be mutually beneficial. A 1987 editorial in the *American Journal of Public Health* sums up this point nicely:

> *The All-Bran story represents a unique and uniquely interesting episode in the history of health education. A private company borrowed a health message from a government agency (with permission), devoted millions of dollars to its dissemination, and thereby increased its sales, while simultaneously conveying the agency's message to the public with a visibility the agency never could have achieved with its own resources.*
>
> Am. J. Pub. Health 77(2): 140–142 (1987)

Still, the FDA needed to do something in response to the Kellogg's campaign. Already, as predicted, the lack of an enforcement action against Kellogg's resulted in many other food companies incorporating health claims in their advertising and labeling. And, as expected, many of these ads and labels presented weak, misleading, and/or unsubstantiated health claim information. Land O'Lakes, for example, was advertising its butter as a good source of Vitamin A, which helps keep your skin soft and smooth. But it failed to point out to consumers the health risks associated with consuming too much fat and the fact that you would need to eat six tablespoons of the butter to obtain the recommended daily allowance of Vitamin A. Despite the growing incidences of abuses, it wasn't until August of 1987 that the FDA finally published its new food health claim

---

[4] Levy, AS and RC Stokes. Effects of a Health Promotion Advertising Campaign on Sales of Ready-to Eat Cereals. *Public Health Rep.*, 102(4): 398–403 (1987).

[5] Freimuth, VS, SL Hammond, JA Stein. Health Advertising: Prevention or Profit. *Am. J. Public Health*, 78(5): 557–567 (1988).

"criteria" in the Federal Register.[6] The FDA would consider the extent to which a health claim on a food product's labeling met these criteria when deciding whether or not the labeling represented "misbranding." The 1987 proposed criteria were as follows:

1. The claim is truthful and not misleading;

2. The claim is supported by valid, reliable, publicly available scientific evidence derived from well-designed and conducted studies consistent with generally accepted scientific procedures and principles performed and evaluated by persons qualified by expertise and training in the appropriate disciplines;

3. That the claim is consistent with generally recognized medical and nutritional principles for a sound total dietary pattern; and

4. The food is labeled in accordance with the existing nutrition labeling requirements.

If the FDA determines that a food meets these requirements, the agency will not consider the food to be a drug within the meaning of 21 U.S. code (USC) 321(p) solely because the labeling contains a health-related message.

As you read these criteria, you may already be able to anticipate some of the problems that the FDA would face when trying to apply and enforce them. There were many subjective and poorly defined elements to these criteria. For example, how will the FDA determine whether the labeling is misleading? What are *"generally recognized medical and nutritional principles for a sound total dietary pattern"*? And perhaps most challenging, how will the FDA determine if studies are valid, reliable, well-designed, and adequately evaluated by persons with appropriate training and expertise? Even the best scientific studies have both strengths and weaknesses. Is it necessary that there be a scientific consensus on the truthfulness of a health claim? Or is it sufficient if some arbitrary numbers of qualified experts agree with the claim?

The 1987 Federal Register announcement included some additional information that would turn out to be perhaps its most controversial aspect. That is, how would the FDA apply these criteria to potential health claims on dietary supplements? Quoting from the pertinent section of the Federal Register announcement[7]:

6 52 FR 28843-28849 (1987).
7 52 FR 28846 (1987).

The agency will apply the same criteria to dietary supplements (products of isolated vitamins, minerals, or other nutrients intended to supplement the diet by increasing total intake of one or more nutrients). However, the extent to which the criteria can be met may be limited. Although supplements are useful for individuals suffering from particular nutrient deficiencies, the available scientific information and data regarding good nutrition and health referred to in this notice focus primarily on the role of traditional foods, not dietary supplements. For example, the National Academy of Sciences (NAS) expressly stated that the findings of its comprehensive study on diet, nutrition, and cancer do not apply to the consumption of dietary supplements.

In effect, the FDA was informing the dietary supplement industry that it would be very unlikely that any of its products would be permitted to make explicit health claims (ostensibly because no dietary supplement will be able to meet FDA's new criteria). The FDA was embarking, once again, on a regulatory strategy that was certain to anger and mobilize the dietary supplement industry and its powerful support base of politicians and consumers.

One of the first important legal challenges to this FDA strategy was from a dietary supplement product known as EXOCHOL, made by Phoenix Laboratories in Hicksville, New York.[8] EXOCHOL was formulated from a variety of phospholipids, primarily lecithin. The labeling and promotion of EXOCHOL asserted that it was "effective in the prevention and treatment of coronary thrombosis, arteriosclerosis, atherosclerosis, and angina," and that it will "prevent cholesterol deposits from forming on the walls of a person's arteries." The FDA argued that this product did not meet the 1987 criteria for health claims, and as a result, it was "misbranded" as an unapproved new drug. The EXOCHOL company argued that the product is a "food for special dietary use" and as such, it is exempt from the drug classification based on the 1987 Health Claim criteria. In its decision, the court found that, based on the history of the labeling and promotion of the product, the FDA could properly classify EXOCHOL as a drug. Recall that based on the FD&C Act's definition of a drug, a product that is "intended" to treat, prevent, cure a disease can still be a food, as well. So the fact that the court determined that the FDA could properly classify EXOCHOL as a drug does not preclude it from

---

[8] U.S. v Undetermined Quantities of an Article of Drug Labeled as "EXOCHOL," US District Court, NY, July 1989.

also being a food. But this was really beside the point in the court's final decision. The EXOCHOL company further argued that their claim of an exemption from the drug definition based on their use of the 1987 Health Claim criteria was no different than what many other companies had done, including the Kellogg's All-Bran ad. On this point, the court agreed. It found that the FDA was not acting "even handedly" in applying the drug definition to products that made therapeutic claims. In its decision, the court stated, "*In every context, the overriding principle of fairness is always the same; the government must govern with an even hand.*" Thus, the FDA lost in its attempt to seize this product as misbranded, not because it was not necessarily misbranded, but because the agency was selectively enforcing the law with some companies and not with others.

Supporters of the dietary supplement industry likely viewed this court case as an important victory. The FDA certainly now realized that it needed a different approach to dealing with unsubstantiated health claims on the labeling of dietary supplements. It was about to get some help from, of all places, the U.S. Congress.

As the 1980s were drawing to a close, health and nutrition professionals, consumer advocacy groups, and the FDA, all recognized that there was a need to improve how nutrition and health information was communicated to the consumer via food labeling and advertising. This recognition went well beyond the issue of "health claims." However, at this time, general nutrition labeling (that is, the inclusion of nutrient and calorie information on food labels) was voluntary, except for specific situations. Food manufacturers were not required to present this information on the labeling of foods unless the food made a nutrient or nutrition claim, or met the legal definition of "foods for special dietary use." If, for example, a food claimed that it was a good source of Vitamin C, or low in fat, or low calorie, etc., then it would trigger the nutrition-labeling requirement. For most other foods, the nutrition labeling was voluntary. But as we discussed earlier, the public was becoming more and more aware of the relationships between diet and health. For food manufacturers, therefore, the nutritional value of foods was now a major promotional characteristic in their marketing strategies. Consumers wanted this information, food manufacturers wanted to provide it (to promote their products), and the FDA needed to make sure that it was presented clearly and accurately.

The FDA and others also recognized that there was a need to standardize the use of *nutrient content claims* on food labeling. Nutrient content claims are label statements that characterize the food as, for example, "low in…" or "reduced…" or "light," etc. The proliferation of the ambiguous and undefined use of these terms on food labels was creating confusion among consumers and unfair competition among food manufacturers. There was an urgent need for the FDA to standardize the use of these terms.

Finally, of course, there was still the issue of health claims on food labeling. The FDA, at this time, was still struggling on how to accommodate and balance the consumers' need and desire for this information, the food industry's desire to take advantage of health claims to market their products, and the agency's responsibility to assure that health claims were clear, accurate, and substantiated by science. Additionally, based on the court's rebuke of the FDA in the EXOCHOL case, the agency had to be sure that any enforcement strategy it adopted would be applied evenly and fairly.

In the late 1980s, the FDA embarked on a series of studies and rulemaking initiatives to address all of these issues. At about the same time, Congress was working toward a legislative solution to the same issues and in the process, incorporated many of the recommendations from the FDA's work. This culminated in the passage of the Nutrition Labeling and Education Act (NLEA) in November of 1990. The NLEA was a landmark piece of legislation and one of the most significant amendments to the FD&C Act since 1938. It addressed all three of the main goals discussed earlier. That is, nutrition labeling would now be mandatory on all food products (with some exceptions), use of nutrient content claims would now be standardized and defined, and the FDA would be given authority to promulgate regulations that would allow disease-related "health claims," if the FDA determined that there was "significant scientific agreement" to support the claim. This specific provision of the NLEA, as passed in 1990, is quoted below.

---

21 USC § 34(r)(3)(B)

The Secretary shall promulgate regulations authorizing claims of the type described in subparagraph (1)(B) only if the Secretary determines, based on the totality of publicly available scientific evidence (including evidence from well-designed studies conducted in a manner which is consistent with generally recognized scientific procedures and principles), that there is significant scientific agreement, among experts qualified by scientific training and experience to evaluate such claims, that the claim is supported by such evidence.

One year after enactment of the NLEA, on November 27, 1991, the FDA published a Proposed Rule in the Federal Register describing how it intended to regulate the Health Claims provisions of the NLEA.[9] Again, it is important to emphasize that the NLEA gave the FDA the authority to establish, by regulation, which health claims would be allowed on foods (see previous box). Any interested party could petition the FDA to have a

---

[9] 56 FR 60537-60856 (1991).

disease-related health claim approved for use, but it would have to pro-vide the FDA with evidence that the proposed new health claim satisfied the "significant scientific agreement" standard. In the Federal Register announcement, the FDA makes it very clear that it will expect a very strong scientific case supporting the claim before it would approve it. In the NLEA, Congress also identified ten initial diet-disease topics that it wanted the FDA to begin evaluating as to their potential for future health claims. These ten topic areas are as follows:

Calcium and Osteoporosis
Sodium and Hypertension
Lipids and Cardiovascular Disease
Lipids and Cancer
Dietary Fiber and Cancer
Dietary Fiber and Cardiovascular Disease
Folic Acid and Neural Tube Defects
Antioxidant Vitamins and Cancer
Zinc and Immune Function in the Elderly
Omega-3 Fatty Acids and Heart Disease

In the FDA's November 1991 Federal Register proposed rules, it announced that it was rejecting the use of health claims for six of these ten, based on the agency's application of the "significant scientific agreement" stan-dard. Specifically, it rejected health claims for the following (the associ-ated Federal Register citations are included in parentheses):

Dietary Fiber and Cancer (56 FR 60656)
Dietary Fiber and Cardiovascular Disease (56 FR 60582)
Folic Acid and Neural Tube Defects (56 FR 60610)
Antioxidant Vitamins and Cancer (56 FR 60624)
Zinc and Immune Function in the Elderly (56 FR 60652)
Omega-3 Fatty Acids and Heart Disease (56 FR 60663)

It is important to note that some of these initially rejected health claims decisions would change in the ensuing years. But this does give you an idea of how strictly the FDA was, at least initially, applying the "signifi-cant scientific agreement" standard. For several years prior to enactment of the NLEA and publication of this proposed rule, many food companies were already using health claims related to these diet-disease topics. But if and when these proposed rules were finalized, those health claims would no longer be permitted. Even the product that started the whole "health claim" storm, Kellogg's All-Bran, would no longer be permitted to make its "fiber and cancer" health claim.

There were several other interesting and noteworthy issues discussed in the FDA's November 1991 Federal Register proposed rule. Some of these would foretell the impending major clash between the FDA and Congress regarding the future regulation of dietary supplements. First, in this Federal Register announcement, the FDA made it clear that it intended to apply the same "significant scientific agreement" standard when approving health claims on dietary supplements as it applies to foods. Even though it did not explicitly suggest that dietary supplements would have a difficult time meeting this standard for any health claim (as it did in the 1987 Federal Register publication), it did, nevertheless, suggest this.

Second, in earlier FDA regulations and FDA-proposed rules related to dietary supplements, it had limited its definition of dietary supplements to only those products composed of essential nutrients (such as vitamins and minerals). Now it was expanding the definition to include "herbs," and suggesting that it may be further expanded to include other products that contain "nonessential nutrient" ingredients.

The third important issue raised in this Federal Register announcement may sound familiar from some of our earlier discussions on this topic. Recall that prior to the passage of the 1976 Vitamin and Mineral Amendment to the FD&C Act, the FDA had attempted to regulate as a "drug" any vitamin and/or mineral supplement that exceeded an upper level of potency. It based this argument on the premise that supplements that provide such high levels of a nutrient serve no "nutritional" purpose, and therefore must be intended for "therapeutic" (drug) purposes. The 1976 Vitamin and Mineral Amendment prohibited the FDA from doing this. In this 1991 Proposed Rule, the FDA was revisiting this strategy with regard to how it would make decisions about health claims on dietary supplements. Consider these passages from the 1991 Federal Register announcement:

For consumption of a substance to have significance within the context of the daily diet, FDA is also proposing ... that the substance must retain its food attributes at the levels that are necessary to justify the claim. For example, if the substance is a vitamin that must be present at a therapeutic level for a health benefit to occur, the supplement would not qualify for a health claim under this proposal. A therapeutic level of a vitamin would be far above that level that is normally characteristic of food, and, consequently, the vitamin would not retain its food attributes. However, FDA is not proposing a specific definition in the general provisions of this proposal for an upper limit of any substance based on the context of the daily diet. Instead, the agency intends to leave it to the petitions

> *that are submitted to demonstrate on a case-by-case basis that the*
> *substance is a food component and is appropriately the subject of a*
> *health claim regulation.*
>
> *FDA is proposing that this provision apply to dietary supplements*
> *as well as conventional foods. ......The proposed provision places*
> *no limits on the potency of safe vitamins and minerals. However,*
> *if a claimed effect can only be achieved at a level of a vitamin,*
> *mineral, or other substance that scientifically cannot be charac-*
> *terized as nutritional, but rather as therapeutic, then that fact*
> *will be considered by the agency in deciding whether the claim is*
> *appropriate for a food, or whether it is in fact a claim that would*
> *make the product a drug under section 201(g)(1)(B) of the act.*
>
> 56 FR 60545-60546

As you can tell from this passage, the FDA is being careful to convey that it is not suggesting that any dietary supplement that provides very high levels of a nutrient (beyond what would be considered "nutrition-ally" relevant), would be regulated as a drug. That, of course, would be in violation of the 1976 Vitamin and Mineral Amendment. Rather, it is sug-gesting that the health claims that are permitted under the NLEA apply to foods, not drugs. And when deciding whether or not the substance that is hoping to make a health claim is a food, the FDA will consider how much of the nutrient in question is in the substance. If the substance con-tains levels of the nutrient that are far in excess of what you would expect of a food (to provide nutritional value), then the FDA would argue that the nutrient is providing a drug-related, therapeutic benefit, rather than a food-related, nutritional benefit. If the FDA was to permit such a health claim, it would essentially be forcing the substance into the "drug" legal definition. By rejecting such a health claim, the substance can still be con-sidered a food, albeit without the benefit of a health claim.

Needless to say, the dietary supplement industry was not pleased with this proposed FDA regulatory approach. They felt that they were being unfairly prevented from taking advantage of the health claims pro-visions of the NLEA to promote the sale of their products. With this pro-posed rule, the FDA was setting the stage for another confrontation with the dietary supplement industry and its political and consumer-based supporters. This latest confrontation would radically alter the way that dietary supplements are labeled, advertised, and sold in the United States to this day.

chapter eight

Congress takes action

In its 1991 Federal Register announcement, the Food and Drug Administration (FDA) made it clear that when deciding on health claims for dietary supplements, it intended to apply the same "significant scientific agreement" standard that it applied for foods. With regard to foods, the FDA had no choice in this matter. The Nutrition Labeling and Education Act (NLEA) explicitly stated that this standard must be used when deciding on the approval or rejection of a health claim. However, when Congress wrote the NLEA, they also included explicit language that allowed the FDA to use a different standard when deciding on the approval or rejection of dietary supplement health claims. Consider the following section from the NLEA:

21 U.S. code (USC) 343(r)(5)(D)

(D) A subparagraph (1)(B) claim made with respect to a dietary supplement of vitamins, minerals, herbs, or other similar nutritional substances shall not be subject to subparagraph (3) but shall be subject to a procedure and standard, respecting the validity of such claim, established by regulation of the Secretary.

Out of context, this section of the law is a bit cryptic. What is it actually saying? "A subparagraph (1)(B) claim" is referring to a "health claim." And "shall not be subject to subparagraph (3)" is referring to the "significant scientific agreement standard." So translating this section into more familiar language, it would read, "A health claim proposed for a dietary supplement does not necessarily need to meet the significant scientific agreement standard. Rather, the FDA can identify an alternative standard for deciding on the appropriateness of dietary supplement health claims."

It is not entirely clear what the intent of Congress was in including this section of the law. Were they expecting the FDA to develop a more lenient standard for dietary supplements? Or were they perhaps intending that the FDA should develop an even stricter standard for dietary supplements? I think the latter is unlikely, as the "significant scientific agreement" standard was already a very high bar to clear. Perhaps the Congress simply could not agree on how to handle health claims on dietary supplements and, therefore, decided to defer to the scientific expertise of the FDA to make this decision. Regardless of the intent, the ultimate consequence of including this section of the law was to grant to the FDA broad authority to decide how to handle health claims on dietary supplements. It ultimately decided to use this authority to apply the "significant scientific agreement" standard to both foods and dietary supplements.

This approach was widely supported by most reputable scientific and health advocacy groups across the country, including the American Medical Association, the American Cancer Society, the American Heart Association, the American Dietetic Association (now known as the Academy of Nutrition and Dietetics), the American Institute for Cancer Research, and the Federation of American Societies for Experimental Biology (FASEB). All of these organizations were more concerned about the proliferation of unsubstantiated and often outright fraudulent claims linked to many dietary supplement products. However, needless to say, the dietary supplement industry was not pleased with the FDA's proposed regulatory approach to health claims on their products. And, as I have mentioned several times in this book, the industry had powerful friends in Congress. The Congressional response to the FDA's 1991 Federal Register proposed rules was to pass the Dietary Supplement Act (DSA) of 1992. This law imposed a 1-year moratorium on implementation of the FDA's regulations pertaining to those sections of the NLEA that deal with health claims on dietary supplements. It also required the FDA to redraft their proposed regulations by June 1993, with the expectation that they would become effective by December 31, 1993. The DSA also mandated an investigation into whether or not the FDA unfairly treated the dietary supplements industry, relative to how it dealt with conventional food manufacturers. The final report of this investigation concluded that there was no discrimination against the dietary supplement industry by the FDA.[1]

On June 18, 1993, the FDA published its new proposed NLEA regulations pertaining to all aspects of the nutrition labeling of dietary supplements, including the use of health claims.[2] As expected, the FDA once again proposed that dietary supplements be regulated identically to conventional foods with regard to nutrition labeling, nutrient content descriptors, and of course, health claims. These new FDA-proposed rules would apply the "significant scientific agreement" standard to health claims for both dietary supplements and conventional foods. Thus, despite the

---

[1] Hegefeld, HA. "Overview of federal regulation of dietary supplements: past, present, and future trends." (2000). https://dash.harvard.edu/handle/1/8846738.
[2] 58 FR 33700-33751 (1993).

DSA's imposed 1-year moratorium, the 1993 proposed regulation of dietary supplements was essentially unchanged from the pre-DSA proposed regulations.

But there was another significant and ultimately more controversial FDA announcement in the same day's issue of the Federal Register (June 18, 1993). This was an Advanced Notice of Proposed Rule (ANPRM) regarding FDA's possible future regulatory strategies for dietary supplements [58 FR 33690-33700]. Recall from earlier discussions in this book that an ANPRM is a way for a regulatory agency (in this case, the FDA) to "test the waters" on how it is thinking about particular issues or regulatory strategies, and to obtain early feedback from affected parties. This feedback could then be considered when drafting future new regulations.

The ANPRM did not focus on the issue of health claims. That issue was addressed in detail in the previously mentioned other Federal Register announcement of that day. Rather, this ANPRM was addressing issues related to the potential health risks associated with some dietary supplements and how the FDA could prevent or reduce these risks to public health and safety. The FDA identified two categories of health risks posed by some dietary supplements, what it termed "direct hazards" and "indirect hazards." Direct hazards are those that present a direct toxic or other negative health affect from a specific supplement or supplement ingredient. Indirect hazards are those that may result when a consumer uses a dietary supplement to treat an illness, rather than seek more effective, traditional medical treatment. To quote from the ANPRM[3]:

> *Indirect hazards may occur if the use of a supplement product delays the diagnosis or treatment of a health disorder. This is a particular concern when exaggerated or unfounded claims are made regarding the benefits of a product in treating or preventing serious diseases, such as cancer and AIDS.*

Many health professionals consider this the more serious problem associated with the enormous explosion in dietary supplement use by the general public. One way to deal with these "indirect hazards" is to strictly enforce the regulations against unsubstantiated health claims. The FDA would certainly argue that this was a primary justification for their use of the "significant scientific agreement" standard for health claims on dietary supplements. But the FDA was also clearly concerned about direct

[3] 58 FR 33692 (1993).

hazards associated with some dietary supplements. The FDA acknowledged that the vast majority of dietary supplements on the market at the time were safe (albeit questionably effective). However, it also recognized that serious health problems have been linked to some dietary supplements in the past. In addition, the rapid expansion of dietary supplements from what were typically simple vitamin/mineral mixtures, to much more complicated formulations that could include amino acids, herbal ingredients, and other non-nutrient substances, presented the potential for many more health risks if left unregulated.

One incident in particular had a major impact on the agency's thinking. In 1989, the FDA became aware of the outbreak of a rare disease, known as Eosinophilia Myalgia Syndrome, or EMS, associated with the intake of a dietary supplement containing the amino acid, L-tryptophan. In the box below is a brief overview of the EMS incident and cause, as presented in the FDA's ANPRM Federal Register announcement.[4]

The outbreak of EMS from the use of L-tryptophan-containing dietary supplements has prompted FDA to reexamine its enforcement posture regarding amino acid containing supplements. EMS is a systemic connective tissue disease characterized by eosinophilia (an increase in one type of the white blood cells), myalgia (severe muscle pain), and cutaneous (skin) and neuromuscular manifestations. This illness, which occurred in epidemic fashion in the United States in the summer and fall of 1989, is associated with the use of dietary supplements containing L-tryptophan. To date, more than 1,500 cases, including 38 deaths, have met the Centers for Disease Control (CDC) case surveillance definition of the disease, although the true incidence of the disorder is thought to be much higher.

FDA first learned about problems with L-tryptophan in 1989, following a report from New Mexico about four cases of an illness manifested by myalgia and eosinophilia, in which the common denominator appeared to be the use of L-tryptophan. FDA subsequently issued a strong public warning on November 11, 1989, to discontinue the use of L-tryptophan. On November 17, 1989, in conjunction with CDC, FDA requested a nationwide recall of all over-the-counter dietary supplements containing 100 mg or

---

[4] 58 FR 33696 (1993).

more of L-tryptophan. The agency also issued an Import Alert to detain all foreign shipments of L-tryptophan. On March 22, 1990, the recall was extended to all marketed products containing added manufactured L-tryptophan because of a case of EMS In a patient consuming less than 100 mg daily. (Products containing added L-tryptophan permitted by § 172.320 were excluded from this recall.) The net effect of the recall and import alert was a ban on the oral supplement forms of L-tryptophan because virtually all of the raw material used to formulate U.S. products was imported.

Despite recent intense research, the exact cause of EMS and an understanding of how it develops have not been established. Initial epidemiological studies implicated the L-tryptophan produced by a single Japanese manufacturer, Shows Denko K. K., and further noted that certain impurities were identifiable in batches of case-associated L-tryptophan. These findings suggested that some impurity or other component in these batches of L-tryptophan may have been responsible for EMS. However, both initial and subsequent epidemiological studies on the EMS epidemic have identified cases of EMS, and another related disease, eosinophilic fascitis, that occurred before the 1989 epidemic and that appear to be related to other batches or sources of L-tryptophan.

EMS and other related disorders are also reported to be associated with exposure to L-5-hydroxytryptophan, a related compound that is not manufactured using the biofermentation process that was used for production of L-tryptophan and is, therefore, not associated with the same impurities or contaminants. There is also some evidence for predisposing factors in some EMS patients. These data, as well as data from animal experiments, indicate that L-tryptophan, either alone or in combination with some other component in the supplement products, may be responsible for some of the pathological features in EMS. Taken together, these findings support previous suggestions that the L-tryptophan-associated EMS was caused by several factors and is not necessarily related to a contaminant in a single source of L-tryptophan.

[58 FR 33696]

In response to the L-tryptophan EMS incident, the FDA commissioned an independent review of the safety of amino acid-containing dietary supplements. The review was conducted by the Life Sciences Research Office of FASEB (LSRO/FASEB). FASEB scientists represent a wide variety of life and health science disciplines. The LSRO/FASEB report was submitted to the FDA in July of 1992. One of the most important conclusions from this report was that the commission "was not able to identify a safe level of intake in dietary supplements for any of the amino acids in the report."[5] This was a very strong conclusion. The LSRO/FASEB commission was considered to be an independent and unbiased scientific group. Their conclusion raised concerns about the safety of any amino acid-containing dietary supplement, not just L-tryptophan.

While the LSRO/FASEB was conducting its review, the FDA also established an internal "Dietary Supplement Task Force" (composed of FDA staff with regulatory, legal, nutritional, and medical expertise) to determine what safety concerns exist with dietary supplements and what regulatory approach should be used to address any safety concerns. A report from this task force was completed in May of 1992. With regard to amino acids in dietary supplements, the task force recommended that amino acid-containing dietary supplements be regulated as drugs, based on its determination that "the primary intended use of these products is for therapeutic rather than nutritional purposes." Alternatively, the task force suggested that amino acids in supplements could be regulated as food additives or Generally Recognized as Safe (GRAS) ingredients, if an upper safe level of use could be identified (as is required for any food additive ingredient). Of course, given the fact that the LSRO/FASEB report would conclude that it "was not able to identify a safe level of intake in dietary supplements for any of the amino acids," this approach would not have likely been feasible.

The suggestion that the FDA was considering regulating amino acid-containing dietary supplements as drugs did not go unnoticed by the dietary supplement industry nor its allies in Congress. Even though this was only an ANPRM in the Federal Register and was not anywhere close to a formal regulatory proposal, the dietary supplement industry seized on the FDA announcement to stoke public outrage at what it saw as FDA's attempt to once again limit the public's free access to the supplements that it wanted. In 1993, the industry even produced a slick television ad, starring the actor Mel Gibson, portraying the FDA as a police raid team converging on Mel Gibson's house in the middle of the night to arrest him

---

[5] 58 FR 33691 (1993).

and confiscate his supply of vitamin supplements.[6] The ad concludes with a voiceover by Mel Gibson:

> *The federal government is actually considering classifying most vitamins and other supplements as drugs. The FDA has already conducted raids on doctors' offices and health food stores. Could raids on individuals be next?*
>
> *If you don't want to lose your vitamins, make the FDA stop. Call the U.S. Senate and tell them that you want to take your vitamins in peace. If enough of us do that, it'll work.*
>
> Mel Gibson.

Of course, the FDA was not proposing to take away the public's free access to vitamins, nor to classify all vitamins and other supplements as drugs. However, it *was considering* classifying amino acid-containing supplements as drugs, and this was enough to set off alarms within the industry. And the industry's scare tactics, along with a carefully orchestrated campaign consisting largely of pre-formatted and addressed postcards to Congress available in health food stores and health food magazines, paid off. The U.S. Congress was inundated with an avalanche of mail opposing the FDA's supposed war on the public's free access to vitamins.

The 1993 ANPRM announcement in the Federal Register discussed other possible regulatory approaches that, if implemented, could have radically altered the regulation of all dietary supplements, not just those containing amino acids. I already alluded to the FDA's consideration of the use of its food additive authority under the Food, Drug, and Cosmetic (FD&C) Act to regulate amino acids as dietary supplement ingredients. However, in addition to amino acids, the FDA was considering this approach to regulate any ingredient in a dietary supplement, including vitamins, minerals, herbs, and other substances. To explain how this would work, we need to first review some of the basic aspects of food additive law and regulation.

In 1958, Congress passed the Food Additive Amendment to the FD&C Act. This amendment created a new legal classification of food substances, officially and legally termed "food additive" and established

---

[6] www.youtube.com/watch?v=IV2olDA0w8U.

a basic framework for their regulation. According to the Food Additive Amendment, a food additive was defined as follows:

21 USC § 321(s)

The term "food additive" means any substance, the intended use of, which results or may reasonably be expected to result, directly or indirectly, in its becoming a component or otherwise affecting the characteristics of any food (including any substance intended for use in producing, manufacturing, packing, processing, preparing, treating, packaging, transporting, or holding food; and including any source of radiation intended for any such use), if such substance is not generally recognized, among experts qualified by scientific training and experience to evaluate its safety, as having been adequately shown through scientific procedures (or, in the case of a substance used in food before January 1, 1958, through either scientific procedures or experience based on common use in food) to be safe under the conditions of its intended use; except that such term does not include—

(1) a pesticide chemical residue in or on a raw agricultural commodity or processed food; or

(2) a pesticide chemical; or

(3) a color additive; or

(4) any substance used in accordance with a sanction or approval granted prior to September 6, 1958, pursuant to this chapter, the Poultry Products Inspection Act [21 U.S.C. 451 et seq.] or the Meat Inspection Act of March 4, 1907, as amended and extended [21 U.S.C. 601 et seq.];

(5) a new animal drug.

There are a number of important points to note about this definition. First, according to the law, a food additive is any substance, "the intended use of which" results in it becoming part of the food or affecting the characteristics of the food. This means that substances that get into foods "unintentionally" (for example, environmental contaminants) are not considered food additives. They would be regulated as a different category of substances with a different regulatory framework. Second, the definition of food additive includes substances that become part of the food "directly or indirectly." So, for example, if a chemical migrates from a food packaging material into the food itself, it would be considered a food additive that got into the food "indirectly." Third, the definition states that if a substance is "generally recognized, among experts qualified by scientific training and experience to evaluate its safety... to be safe," then it is not, by legal definition, a food additive. This is the legal derivation of the so-called GRAS substance. GRAS substances make up the vast majority of substances and ingredients in our food supply. They are not, by legal definition, food additives and they are regulated differently. We will discuss GRAS substances in a bit more detail later. Fourth, the definition includes a list of additional substances that may be found in foods that, like GRAS substances, are not legally considered food additives. These include pesticide residues, color additives, substances that were already regulated before passage of the 1958 Food Additive Amendment (so-called prior sanctioned substances), and animal drugs in foods. All of these nonfood additive substances are regulated in foods by different regulatory procedures.

The FD&C Act goes on to state

21 USC §342 Adulterated food

A food shall be deemed to be adulterated—
(C) if it is or if it bears or contains (i) any food additive that is unsafe within the meaning of section 348 of this title.

In other words, if the food additive is unsafe (as described in Section 348 of the law), the food would be considered "adulterated" and would be in violation of the FD&C Act. So, how does Section 348 of the FD&C Act address the safety of food additives? In order for the FDA to approve a new food additive for use in foods, the manufacturer would be required to submit a petition to the FDA requesting that the substance be approved. The 21 USC §348 goes on to specify what would need to be included in that petition.

21 USC §348(b)

(1) Any person may, with respect to any intended use of a food additive, file with the Secretary a petition proposing the issuance of a regulation prescribing the conditions under which such additive may be safely used.

(2) Such petition shall, in addition to any explanatory or supporting data, contain—

(A) the name and all pertinent information concerning such food additive, including, where available, its chemical identity and composition;
(B) a statement of the conditions of the proposed use of such additive, including all directions, recommendations, and suggestions proposed for the use of such additive, and including specimens of its proposed labeling;
(C) all relevant data bearing on the physical or other technical effect such additive is intended to produce, and the quantity of such additive required to produce such effect;
(D) a description of practicable methods for determining the quantity of such additive in or on food, and any substance formed in or on food, because of its use; and
(E) full reports of investigations made with respect to the safety for use of such additive, including full information as to the methods and controls used in conducting such investigations.

There are three important and relevant points to note regarding this section of the law. First, in order for a new food additive to be approved by the FDA, the manufacturer must provide information on what it is intended to do in the food (its function). The FDA would not approve a new food additive that serves no function in the food. Second, the manufacturer must provide data on how much of the substance is needed to accomplish its function in the food. The FDA would only approve the use of the food additive up to the amount needed to accomplish its function in the food. And third, the manufacturer must provide full reports of its safety/toxicity testing of the substance. This third requirement can easily cost the company many millions of dollars in laboratory and animal testing. This may be worthwhile for a company that is seeking approval for a novel food additive (say, for example, a new artificial sweetener) for which it could obtain patent protection while it recouped the costs of the food additive testing and still make a profit. But a dietary supplement manufacturer would not be able to obtain patent protection for most dietary supplement ingredients (a vitamin, mineral, amino acid, herb, omega-3 fatty acid, etc.). If the FDA approved the ingredient as a food additive, any competing dietary supplement manufacturer could use the newly approved food additive ingredient in their dietary supplement. This would effectively deny the original petitioning company the opportunity to recoup its testing costs and make a reasonable profit through patent protection.

If the FDA was ever to use its food additive authority to regulated ingredients in dietary supplements, it would not likely apply this approach to vitamin or mineral ingredients. Many of these vitamins and minerals were already considered GRAS and thus did not fit within the legal definition of a food additive. But what if the amount of the vitamin or mineral ingredient in the dietary supplement, for which it has GRAS approval, exceeds the level needed to accomplish its technical function in the dietary supplement? And what is the "technical function" of a vitamin or mineral ingredient? For dietary supplements, the FDA would consider the technical function of a vitamin or mineral to be to provide *nutritive value*. If the manufacturer adds very large amounts of a vitamin to a supplement, is that amount still serving its purported function of providing nutritive value? Or is that high amount of the vitamin serving some other function (i.e., therapeutic) that is not part of the GRAS use approval of the vitamin? Before we consider these questions, let's take a closer look at the legal meaning of GRAS substances.

Neal Fortin, in his book, *Food Regulation: Law, Science, Policy, and Practice* (2nd Ed. 2017) does an excellent job of presenting a simple and clear explanation of this category of food substances. One of the most important points that he makes, and one that is certainly relevant to the issue of dietary supplement ingredients, is that the term, *GRAS substance*, does not accurately describe this category of food substances. The more correct term is *GRAS use of a substance*.[7] This is because granting GRAS status pertains to the particular use of the substance, and not to the substance itself. The example Fortin uses to illustrate this distinction is caffeine. He notes that caffeine is not a GRAS substance. Rather the use, for example, of caffeine in colas up to 0.02% is recognized as an approved GRAS use of caffeine.

Another important point that Fortin notes is that determining whether the use of a particular substance is truly "GRAS" is not a simple or trivial process. As the law explicitly states, it requires that its safety be "generally recognized among experts qualified by scientific training and experience to evaluate its safety, as having been adequately shown through scientific procedures... to be safe under the conditions of its intended use." In other words, there needs to be good scientific data supporting the safe use of the substance and that these data must be "generally recognized" among scientific experts as having established the safe use of the substance. "Generally recognized" is a high standard to meet, requiring a very strong consensus among the scientific experts. Still, the procedure for obtaining GRAS use status for a substance is certainly less complicated and less costly than the new food additive petition process described earlier.

---

[7] Food Regulation: Law, Science, Policy, and Practice. By Neal D. Fortin. John Wiley & Sons, Hoboken, NJ. 2nd Ed. (2017).

In addition, consider the following section of the Code of Federal Regulations (CFR) pertaining to GRAS use of a substance:

21 CFR §170.30

(h) A food ingredient that is listed as GRAS in part 182 of this chapter... shall be regarded as GRAS only if, in addition to all the requirements in the applicable regulation, it also meets all of the following requirements:

(2) It performs an appropriate function in the food or food-contact article in which it is used.

(3) It is used at a level no higher than necessary to achieve its intended purpose in that food...

So, if the "appropriate function" for the GRAS use of a vitamin or mineral ingredient in a dietary supplement is to provide "nutritive value," does that vitamin or mineral ingredient still qualify for GRAS use status if the amount of the vitamin or mineral exceeds that needed to provide "nutritive value?" We will revisit this question shortly.

From 1958 (after passage of the Food Additive Amendment) through 1973, the FDA maintained a list of GRAS uses of various substances. After 1973, FDA established a "GRAS affirmation petition" process, by which food manufacturers could seek GRAS use approval for a substance from the FDA. But the FDA quickly fell behind in making decisions on these petitions. So, in 1997, it established a more streamlined GRAS use notification process, by which companies could submit a report with supporting data to the FDA. The FDA would then simply respond with either a "no questions" determination (meaning that the FDA agrees with the company's self-determination of GRAS status) or respond with an "insufficient basis" determination (meaning that the report from the company did not provide enough information on which to base a GRAS use approval). The FDA maintains a web-based list of these GRAS notifications, knows as the GRAS Notice Inventory. It can be found at the following web address: www.fda.gov/Food/IngredientsPackagingLabeling/GRAS/NoticeInventory/default.htm

It is interesting and informative to look at some of the reports that companies have submitted as part of this inventory list. These reports are typically quite long and include extensive information on how the substance will be used in foods and data supporting the safety of the substance at its intended use levels.

So how does all of this relate to the FDA and its ANPRM of June 1993? As mentioned earlier in this ANPRM, the FDA was already considering using its food additive and GRAS statutory authority to regulate the use of amino acids in dietary supplements. Based on this authority, the manufacturer would need to prove to the FDA that the levels of amino acids they want to add to their supplements are safe (within the meaning of the requirements for either food additives or GRAS use, as just previously discussed). But the "scientific experts" that would be needed to make this determination would be very similar to those who made up the LSRO/FASEB Commission mentioned earlier. And this commission had already concluded that they were not able to identify a safe level of intake for amino acid-containing dietary supplements. So, the FDA was effectively informing the supplement industry that any food additive petition or GRAS use notification for an amino acid ingredient would likely be rejected by the agency due to safety concerns.

How does this relate to other dietary supplement ingredients (other than amino acids)? Let's start by again considering simple vitamin and mineral ingredients. Any ingredient in any food must fall into a specific legal category of food substances. In the case of vitamins or minerals in dietary supplements, the only options are to either classify them as food additives or GRAS use substances. If they are to be classified as food additives, they would need to go through the new food additive petition process. As outlined earlier, this is a very costly and slow process. But most of these vitamins and minerals were already listed as GRAS use substances. They were either listed or affirmed as GRAS by the FDA (as listed in 21 CFR Parts 182 or 184) or independently determined to be GRAS by the manufacturer without officially submitting a GRAS notification. This "independent determination" of GRAS status is not new. Many substances in our foods are considered so-called "unlisted GRAS use substances." These could include anything from table sugar to the pieces of potato in a can of beef stew. These substances are so obviously "safe" that there is no need to notify the FDA of their use in foods, yet they are still considered GRAS. But the FDA could always revisit their GRAS status and require that it be affirmed if new information becomes available regarding their use or safety. As discussed in the ANPRM, one way that they could apply this to the regulation of vitamins or minerals in dietary supplements would be to "affirm as GRAS... the highest RDA levels listed by the National Academy of Sciences [for the particular vitamin or mineral]."[8] If manufacturers wanted to add higher levels of the vitamin or mineral to the supplement (beyond that recommended by

[8] 58 FR 33695 (1993).

the National Academy of Sciences to meet nutritional needs), they would have to "submit evidence that would justify a higher level that represents safe use."[9]

It could be very tricky for a company to try to do this. We know from our previous discussions in this book that dietary supplement companies want to ensure that their products are regulated as foods, and not as drugs. In the Starch Blocker case (Nutrilab Inc v. Schweiker, 1983) the court found that in determining whether a product is a food or a drug, consideration should be given to how it is consumed. If it is consumed primarily for taste, aroma, or nutritive value, it is a food and not a drug. Dietary supplements are clearly not consumed for taste or aroma. But they are presumably consumed for "nutritive value." If a dietary supplement company is required to justify adding levels of a vitamin to a supplement higher than that recommended by the National Academy of Sciences (as part of its affirmed GRAS use status), then the FDA could argue that these higher levels go beyond what is needed for "nutritive value." The result may be that the supplement no longer meets the legal definition of a food. And if they are not a food "intended to alter the structure or function of the body," then they are a drug. One might consider this regulatory approach to be a bit of a "legal trap" being set by the FDA to get around the restrictions imposed on the agency by the 1973 Vitamin and Mineral Amendment to the FD&C Act.

This issue becomes even more complicated when you consider dietary supplements that contain other "non-nutrient" ingredients, such as herbs, extracts, or other substances. At the time of the ANPRM (1993), many of these substances were approved food additives or had approved GRAS use status. However, the basis for their approval was for various technical uses in the food. For example, many of these substances were approved for use as stabilizers, flavoring agents, firming agents, emulsifiers, etc. As the FDA noted in its ANPRM, "food-use herbs are subject to the food additive provisions of the [FD&C] act." Yet many of these substances were already being extensively used as ingredients in dietary supplements, without any safety review for this purpose, "based, presumably, on a GRAS determination by the marketer." If the FDA were to decide to pursue the food additive provisions of the law to regulate these ingredients (as they were hinting to in the ANPRM), then the manufacturer would have to embark on the very costly and time-consuming process of submitting a new food additive petition for each ingredient. In addition to providing the extensive safety/toxicity data required of this petition process, the food additive law also requires the following:

[9] Ibid.

21 USC 3§48(c)(4)

If, in the judgment of the Secretary, based upon a fair evaluation of the data before him, a tolerance limitation is required in order to assure that the proposed use of an additive will be safe, the Secretary—

(A) shall not fix such tolerance limitation at a level higher than he finds to be reasonably required to accomplish the physical or other technical effect for which such additive is intended;

In other words, the manufacturer can only add a level of the ingredient (herb, extract, etc.) up to the amount needed to accomplish it purpose in the food (in this case, the dietary supplement). But what is the "physical or other technical" purpose of these ingredients in a dietary supplement? Unlike the vitamin and mineral ingredients discussed previously, these herbal and other ingredients are not nutrients per se and thus the FDA could argue that they do not provide "nutritive value." Once again, if the supplement does not provide "taste, aroma, or nutritive value," and the supplement is intended to "alter the structure or function of the body," does it not then meet the legal definition of a drug?

Clearly, the dietary supplement industry had reason to be nervous regarding the ideas raised by the FDA in the 1993 ANPRM. These were ideas that go far beyond health claims and whether or not the FDA should apply the "significant scientific agreement" standard to health claims on dietary supplements. These ideas, if ever formalized into federal regulations, would have radically altered how dietary supplements could be formulated and marketed. But even before the June 1993 ANPRM, the Congress was already moving toward a major legislative intervention to protect the dietary supplement industry from further regulatory intrusion by the FDA. On April 7, 1993, Senator Orrin Hatch introduced S. 784 in the Senate, and Congressman Bill Richardson introduced the companion bill, H.R. 1709 in the House. Both bills were designed to provide the dietary supplement industry with the ability to more freely manufacture, formulate, label, and advertise their products, and significantly restrict the power of the FDA to regulate these products. By the summer of 1993, shortly after the Federal Register ANPRM, hearings on the two bills were underway in the House and the Senate. One of the most interesting and contentious hearings was the Senate Committee on Labor and Human Resources hearing on October 21, 1993, titled, "Legislative Issues Relating to the Regulation of Dietary Supplements." Senator Orrin Hatch, the dietary supplement industry's most powerful ally in Congress, was a member of this committee. The FDA Commissioner at the time was Dr. David Kessler. Dr. Kessler came to the FDA in 1990, uniquely qualified for a leadership position in public health policy, having earned both a medical degree from Harvard University as well as a law degree from the University of Chicago. Even before these dietary supplement hearings, Dr. Kessler had already built quite a reputation for himself as FDA Commissioner for his tough stance against the tobacco industry. But he was also recognized for his efforts to bring attention to what he saw as widespread misinformation and even health fraud in some parts of the dietary supplement industry. It was, after all, under his leadership that the FDA steadfastly refused to weaken the "significant scientific agreement" standard for health claims on dietary supplements. And it was under his leadership that the ideas in the June 1993 ANPRM were put forward. So it should come as no surprise that the interaction between Senator Hatch and Dr. Kessler during the October 1993 Committee on Labor and Human Resources hearings would be heated and contentious.

chapter nine

The congressional hearings

Legislative Issues Relating to the Regulation of
Dietary Supplements

*Hearing Before the Subcommittee on
Labor and Human Resources
United States Senate
October 21, 1993*

As I mentioned in the previous chapter, a number of hearings were held
in Congress relating to the House bill (H.R. 1709) as well as the Senate bill
(S. 784). The Senate Subcommittee on Labor and Human Resources held
one of the most interesting hearings in October of 1993. In this chapter,
I am going to present excerpts from the transcripts of this hearing. I am
also providing links to the complete hearing transcript and to a video of
excerpts from the hearing.[1] Both the complete transcript and the video
recording are very interesting to read and view. They provide much more
detail from the hearing than I can include in this chapter. And the video
gives you a much better sense of the tone of the hearing and the attitudes
of the hearing participants. Nevertheless, I think it is interesting and
worthwhile presenting some excerpts from the hearing in this chapter.
I will only include some minor comments and will instead let the transcript speak for itself. Many of the issues that I have already discussed
in previous chapters of this book are addressed, in various ways, in the
excerpts that I have selected. As such, I think that they offer a particularly
relevant context in which to consider the factors that ultimately led to the
passage of the Dietary Supplement Health and Education Act of 1994. I am
also including some screen capture images of some of the hearing participants, taken from the actual video recording of the hearing.

Following some opening comments from Subcommittee Chair
Senator Ted Kennedy (D-Massachusetts), Senator Orrin Hatch (R-Utah)
made an opening statement. Recall that Senator Hatch was (and still is) a
longtime supporter of the dietary supplement industry and the principle
sponsor of the Senate bill, S. 784.

---

[1] Transcript of Hearing: www.uvm.edu/~spintaur/Book/Hearing.pdf; Video excerpts
from Hearing: www.uvm.edu/~spintaur/Book/Hearings/Hearings.html.

**Senator Orrin
Hatch
(R-Utah)**

Opening Statement

Well, thank you, Mr. Chairman, and thank you. Senator Kassebaum, and thanks to Congressman Richardson and Congressman Gallegly, our leaders in the House on this issue.

We stand at a crossroads today. Either we can choose to move forward: We can resolve the issue of dietary supplements. We can move S. 784 to the House. And we can allow consumers what they demand—free access to safe dietary supplements.

Or we can double back: We can choose to do nothing. We can allow the Food and Drug Administration (FDA) to continue its life and death grip on products which have been proven to enhance public health. We can watch the moratorium on health claims expire and wait for up to 100 million angry American citizens to descend.

Frankly, I would not want to be in Washington if we allow the latter to happen. [Laughter.]

It is no secret to anyone in this room how I feel about dietary supplements. I really believe in them. I use them daily. They make me feel better, as they make millions of Americans feel better. And I hope they give me that little added edge as we work around here.

And if anyone in the room doubts that, just check with your mailroom. I think they will tell you the real story. It is no secret that the dietary supplement industry is large in Utah, some $700 million to $1 billion a year.

And it is no secret that Bill Richardson, Elton Gallegly, and I are leading the army of citizen protestors for whom we drafted the Dietary Supplement Health and Education Act, S. 784 and H.R. 1709. Congressmen Richardson and Gallegly have done a great job on this, and I, of course, appreciated Congressman Richardson's testimony this afternoon.

But you, Mr. Chairman, and you, Senator Kassebaum, are not getting the credit you deserve. You are the generals in this army. Your efforts behind the scenes are leading us toward passage of a bill, toward a victory for free choice.

And I want to recognize publicly today the commitment to resolving the issue that you both have demonstrated. It is the only way we are going to end the "Vitamin Wars," as I believe we will. This is a tremendously complicated legal issue which can be expressed very simply: Are we in Congress going to allow one tiny agency to restrict the access of millions of Americans to safe products they wish to use to improve their health? Are we in Congress going to elevate to red-button priority our national dialog on health care reform, then turn around and allow one misguided cadre of bureaucrats to restrict the information consumers need to be more healthy?

Clearly, 59 Members of the Senate, 59 bipartisan Members of the Senate, and over half of this committee have responded with a resounding "No." ...

And three times as many House members have joined with Bill Richardson and Elton Gallegly to halt this nonsense. "Nonsense" is the polite term for what I see happening at the FDA. Let me say for the record what I have told many of you privately. I have the greatest admiration for Commissioner Kessler. He is an honorable man. He has worked with many of us, and he is first rate. And I think you would search long and hard to find any Member of Congress who is a bigger fan of the FDA than I. But on this issue, the FDA is simply wrong. I want it to be clear. Congress is on the record as saying so. There are two primary issues that have prompted this great consumer outcry over FDA's treatment of dietary supplement. One is access; the other is claims.

On access, the FDA has used tortured legal authorities to try to remove dietary supplements from the marketplace. Some of these products were never alleged to be unsafe, much less proven to be unsafe. As you will hear later, and as Congressman Richardson said, the court has termed FDA's actions "Alice in Wonderland." And I, for one, am tired of the tea party.

On claims, the FDA has used a jumbled-up process and a strict interpretation of a good law, one that Senator Metzenbaum and I and many others worked to put through, to block most consumers from receiving truthful and scientifically accurate information about dietary supplements.

The FDA has come to Congress decrying "snake oil." The Commissioner has testified before the House that "we are back at the turn of the century, when snake oil salesmen could hawk their potions with promises that couldn't be kept." The Commissioner's deputies have hit the airwaves, holding up products you are sure to see later, showing claims they believe to be false and misleading.
And I have a simple question. If the FDA feels there is a problem, why doesn't it remove them from the marketplace and protect the public rather than condemn the product of the house on TV?

One month, the FDA sends a report to Congress stating, "The vast majority of dietary supplements consumed today do not raise serious health or regulatory concerns." The next month, the Commissioner testified before the House that "for every dietary supplement in the marketplace that may have some value, there are 100 or 1,000 that are worthless." The FDA says that 80% of the market is safe and of no concern. By FDA's own estimate, there are only about 40,000 products. If 1 per 100 or 1 per 1,000 is bad, how can 80% be safe? I sure would not want the FDA keeping my checkbook. [Laughter.]

The FDA has presented the Congress with a report, "Unsubstantiated Claims and Documented Health Hazards in the Dietary Supplement Marketplace." This false and misleading document is so riddled with inaccuracies that it lacks any evidentiary value and raises serious questions about the motives of those who are responsible for its preparation. There can be little doubt that the report was hastily thrown together for a dramatic unveiling at a House hearing. The FDA completely ignored the rigorous preapproval requirements for surveys in the Paperwork Revolution Act... It should be Revolution Act. [Laughter.] According to the agency's own internal documents, barely 3 weeks before the July 29 hearing, 63 agency officials were sent out in a nationwide "undercover survey" to find examples of health food store employees making unsubstantiated claims. Their operating instructions were laid out in a "not for public distribution" memorandum, which instructed these employees on how to dress and act, what leading questions should be asked, and they were cautioned not to discuss the assignment outside of the office.

The results of their investigation were to be conveyed secretly on a specially prepared reporting form. It is hard to imagine a clearer case of Government entrapment and misuse of taxpayer dollars. I will just read a short excerpt from one of these cloak-and-dagger field reports:

"The Consumer Safety Officer entered the store as a person off the street. She walked through the aisles to the east wall, where various nutritional supplements were displayed on shelves." "She was approached by a store employee (white female, blond hair, approximately five feet five inches) who asked if she could help the CSO." Maybe we should do a bill to merge the FDA with the FBI.

The FDA has participated in—notice I did not say conducted—at least one armed raid of a medical doctor who was dispensing supplements. The FDA has *gone on television assuaging the public that it does not want to remove supplements from the marketplace, yet has issued Federal Register proposals revealing an intent to restrict amino acids, herbal products, and high-potency vitamins.*

Today, I am releasing a report prepared by my staff titled, "False and Misleading: FDA's Report: Unsubstantiated Claims and Documented Health Hazards in the Dietary Marketplace." Based on this analysis, it is clear that the FDA report is so riddled with inaccuracies that it lacks any evidentiary value and raises serious questions about the motives of those who are responsible. I am asking the Clinton administration today to withdraw the report and to take the necessary steps to make sure in the future information provided to Congress and the American people is both accurate and unbiased and is gathered pursuant to Federal law.

The FDA has taken a solid law, the Nutrition Labeling and Education Act (NLEA), and turned it on its head. The point of the NLEA was to educate the consumer about good nutrition, not to block information. I was a cosponsor of the NLEA, but my work on the legislation pales in comparison to your legendary efforts, Mr. Chairman, and those of Senator Metzenbaum. I remember when the bill was on the floor and Senator Metzenbaum and I worked so diligently to make sure certain dietary supplements were treated properly. I remember when we talked about wanting to provide the public with better nutrition information and wanting to enable consumers to select foods to protect and improve their health.

I remember when Senator Metzenbaum said that whatever approach the Secretary of HHS chose to take on supplement labeling, the "system must be based on the same considerations that guide other agency decisions: public health, sound scientific principles, and consumer fraud." Who could disagree with that? The FDA could, that is who.

Is FDA protecting the public health by refusing to allow pregnant women to be informed that 0.4 mg of folic acid taken daily could dramatically reduce their chance of having a baby with birth defects? Is FDA protecting the public health when 100 babies are born a month whose birth defects could have been prevented? Is FDA protecting the public health by holding up the shield of "significant scientific agreement" to block all supplements but calcium from bearing health claims?

Is FDA protecting the public health when it turns down a health claim for antioxidants, even though surveys show eight out of ten doctors regularly use Vitamin E?

Before I close, Mr. Chairman, I would like to recognize several outstanding people who have worked closely with us. Our schedule did not allow them to testify, but I want to make sure that their statements are included in the record, along with a number of others I will submit.

Finally, Mr. Chairman, I want to thank you for scheduling this hearing. I know it was not easy. Your plate is extremely full, and especially at this time of the year, and your capable staff has been most helpful to us. I want to thank you, too, Senator Kassebaum, for your generous offer to work with us to effect a solution. I believe your influence on this process clearly has been felt.

I say to all here, in all sincerity, that I want to resolve this issue. I believe we will do so. I recognize the concerns many, even cosponsors, have expressed that S. 784 does not adequately address the safety issue. I recognize concerns that the language is not drawn tightly enough to prevent false and misleading claims. I recognize concerns over setting up a dual standard under the NLEA for foods and for supplements. I want to resolve all these concerns. I do not intend that we allow "snake oil" to be marketed or that we allow unsafe products on the market or that foods be treated unfairly.

What I do intend, Mr. Chairman, is to allow consumers access to safe products and to information about those products. What I do intend is to stop FDA's regulatory over-reach and allow the agency to focus on real problems, such as medical devices. And, finally, what I do intend, Mr. Chairman, is to get a bill to the President as quickly as possible. I hope that we can count on you for your support. There are a number of groups and individuals who have worked closely with us during our consideration of this issue. I would like to recognize those groups and individuals and submit their statements for the record: Citizens for Health, a grass-roots consumer organization; and the Alternative Treatment Committee of the AIDS Coalition to Unleash Power, ACT-UP, if you will, in San Francisco, both of which have provided me with tremendous amounts of assistance and help. Their representatives were not able to be here with us today but would like to submit their statements for the record.

Another individual who has been a great help to me is Dr. Julian Whitaker, a noted physician, author of the monthly newsletter "Health and Healing," and president of the American Preventive Medical Association. He has provided a statement titled "Regulation of Dietary Supplements."

In the audience today is Claire Farr, president of Claire Industries in San Marcos, California, who will submit information for the record on amino acids. And I would also like to recognize Dr. John C. Godfrey and Dr. Robert Pollock who are with us today. They are two of the pre-eminent researchers on amino acids in the scientific community today. I am submitting their statement for the record.

Also in the audience, Dr. Alvin B. Siegelman from my home State of Utah, vice president of corporate health sciences at Nature's Sunshine Products. Dr. Dennis Jones, an internationally recognized researcher on herbal products, has provided me with information that I also ask to be made part of the record.

> *I want to thank all of these individuals and the many, many others who have worked with us so closely in development of S. 784 and its House counterpart, and I want to thank you again, Mr. Chairman, and you, Senator Kassebaum, for allowing me the time to express my viewpoints on this issue.*

There are a number of interesting points in Senator Hatch's opening statement. First, he makes no apology for the fact that he represents a state with a substantial dietary supplement industry presence (although he does not mention the extent of the financial support he receives from this industry). He is obviously very proud to do whatever he can to support this important industry in his state. Second, Senator Hatch either sincerely believed, or was stoking public fears when he suggested that the FDA was committed to preventing, or significantly limiting, the public's free access to dietary supplements. And third, he was clearly frustrated by the FDA's very rigid stance on approving health claims for dietary supplements, even referring to the "significant scientific agreement" standard controversy. Some of these same concerns are explicitly addressed in Dr. David Kessler's opening statement.

Dr. David A. Kessler

Commissioner

FDA (1990–1997)

Opening Statement

Thank you, Mr. Chairman and members of this committee. First, may I ask that my written statement be included for the record, and also Dr. Philip Lee, who is the assistant secretary of health, if we can submit a statement from him. Dr. Lee is in Houston, and he is very committed to working with this committee to resolve this very important issue.

What I need to do today, what I want to do today, is two things: One, to try to set the record straight; and, two, to pledge to you, this entire committee, that we stand ready to work with you to resolve this very important issue. First, in setting the record straight, the industry's message is simple. It says: Write to Congress today or kiss your supplements good-bye. It is one of the reasons—it is not the only reason—that you are getting a lot of mail. This message is absolutely false. We hear people claiming that FDA is trying to deny consumers the right to take vitamins and minerals or force them to go to a doctor to get a prescription for their Vitamin C. Nothing could be further from the truth.

Mr. Chairman and members of this committee, I am here to reaffirm FDA's commitment to maintaining the American consumer's access to dietary supplements. I support access to dietary supplements, and you should support that access, too. I am here to pledge that the FDA will work with this committee to achieve the goal to which I believe we all can aspire: guaranteed access to a wide variety of supplements that consumers can trust, are safe, and are properly labeled.

Some say we are trying to put the health food industry out of business that products will have to stop being sold. That is simply not correct. Any nutritional supplement currently on the market can be sold as long as it presents no safety problems. As long as these products are safe, manufacturers are not going to run into any problems from the FDA. But there is a point at which I need to draw a line. It is the point at which one of those products on store shelves makes the claim that something is useful in treating diseases such as cancer, diabetes, or arthritis when it, in fact, does no such thing. Manufacturers do not have to take the product off the market. They simply have to remove the unproven claim from the labeling or any promotional material. Sell whatever safe product you will, but do not say that it will prevent, cure, or treat a disease unless you have established affirmatively that it really will.

I have no problems with consumers taking supplements to improve their diets, but when supplements are really drugs in disguise, promoted to treat serious diseases, then I believe we have a problem. Recognize at the outset that the dietary supplement industry is essentially unregulated. When consumers pick up a dietary supplement today, they assume that the product is safe.

But the fact is that there has never been a systematic evaluation of the safety of dietary supplements. And when consumers see a health claim for a dietary supplement, they assume it will provide the benefit it touts. In fact, the marketplace is full of unsubstantiated claims.

Congress sets the standard for health claims for foods in the Nutritional Labeling and Education Act. But you could not reach agreement on the standard for dietary supplements and asked the FDA to set that standard. In November 1991, we proposed that dietary supplements should be subject to the same standard for health claims that you articulated so clearly for foods, not the standard for drugs, for foods; namely, that the claim be supported by significant scientific agreement. We did not see why a health claim should be allowable for a Vitamin C tablet but not for the Vitamin C in broccoli or orange juice. We reaffirmed that position in a proposal we issued this past June.

Make no mistake, Mr. Chairman. There are many supplements on store shelves today making unsubstantiated health claims. The promotion of these products for serious health problems is a real problem. We issued a report, as Senator Hatch said, prior to Congressman Waxman's dietary supplement hearing on July 29. It listed hundreds of products that claim to cure, treat, or reduce the risk of cancer, AIDS, diabetes, heart disease, arthritis, and other diseases. These claims appear in current catalogs, brochures, other advertising materials, and right on the label in certain instances.

In the absence of a clear standard, the best the FDA can do to try to separate the good from the bad when it comes to dietary supplements is to go after products one by one. If there are people who want to go out and buy products such as this one—this one is named Nature's Response—let them do it. But no one should attach to this product this brochure, making the claim that Nature's Response inhibits reproduction of the HIV virus and inhibits the growth of cancer. Increasingly, scientists are uncovering important relationships between diet and health. But in the dietary supplement marketplace filled with unsubstantiated claims, for every legitimate product that may be of some value, there are many that are worthless.

Some exciting advances and scientific advances are being made, but unless something changes, products that provide real benefits—and there are products that provide real benefits—will be drowned out by the hundreds of other products, making unsubstantiated and sometimes downright fraudulent claims. Congress confronted the issue in the 1990 under NLEA. Let's go back to what was happening in the supermarket before you passed NLEA. We had a proliferation of misleading claims and unfounded claims on food packages that undermined consumer's faith in the food label. When the marketplace is flooded with these products making unsubstantiated claims, the products that offer legitimate benefits are lost in the morass of those that offer nothing.

The food industry recognized it had a problem back then. The NLEA was a commitment to restoring credibility on the supermarket shelves. You, Congress, set a standard in NLEA that said preliminary, premature evidence was not an adequate basis for wholesale changes in the diets. You set the standard, significant scientific agreement. Today, under current law, we have a standard for drugs, and we have a standard for foods. We believe that the standard you have already established for foods should be the standard for dietary supplements. Some would have you create a standard for dietary supplements that is weaker than the standard for foods. But the implications of a weaker standard for dietary supplements deserve your full attention. Is weakening the standard what you really want to do at a time when millions of Americans, so many Americans, are taking supplements?

Believe me, I appreciate the promise of a simple cure. Of course, we would all rather take a miracle pill than undergo more arduous and sometimes uncertain treatments. But, unfortunately, cures do not always come packaged as neatly as we would hope.

And patients who forsake therapies that offer some real benefit for the siren song of empty promises have a lot to lose. Some would have you permit marketers to decide whether a health claim is appropriate without review by FDA. Remember that the proliferation of health claims on food labels in the 1980s occurred precisely because companies, rather than FDA, decided what claims could be made. This approach opens the floodgates to claims that have no scientific basis. It puts the consumer in an impossible situation, because there is no way of telling what works from what does not work. Furthermore, if companies are allowed to make claims without sound studies to back them up, there is no incentive to do those studies that will finally determine which products offer real benefits. It would be a sad loss if consumers were to turn their backs on all dietary supplements because their faith was undermined by the proliferation of misleading claims.

But there may be a higher price for consumers, a price greater than the cost of being victimized by worthless products or foregoing therapies with demonstrated usefulness. You know, there is a widespread perception that because something is natural it is safe. We have learned, sometimes the hard way, that equating safety with natural can be a costly equation. Think about it. Half our prescription drugs in this country are derived from plants, and no one doubts for a minute that drugs can have toxic effects. Why, then, should we assume that all risks disappear when plants are sold as dietary supplements for therapeutic purposes?

Dietary supplements have been linked to death, kidney and liver failure, nerve damage, psychosis, and we have only discovered it after the injuries have taken place. How can we work together to ensure the availability of safe dietary supplements that do not cross the line separating legitimate claims from bogus assertions?

We would ask you to consider doing three things:

First, guarantee access to dietary supplements. Write your constituents and tell them that Congress is working with the FDA to maintain access to dietary supplements. I guarantee that access.

Second, give the FDA an effective means to deal with demonstrable safety problems and to ensure that these products are properly manufactured.

And, third, hold manufacturers of dietary supplements to the same standard that you previously established for health claims on foods. Support that standard. Do not lower it. In the end, this approach will benefit consumers because it will keep insupportable health claims off the shelves, and it will give consumers access to meaningful choices, choices based on science and not salesmanship.

I know, when it comes to dietary supplements, the emotions are running very high. The time has come to lower the emotional tenor of that debate. And I know you are receiving more mail from constituents on this issue than seems imaginable. The time has come to stem the flood of letters by acting decisively and responsibly to support access to dietary supplements, but not to allow unsubstantiated claims to be made for those products. It is time to sit down, to work things out, and to find a solution.

Thank you, Mr. Chairman.

Dr. Kessler's message in his opening statement was fairly straightforward and direct. He had two main points to make. First, he was trying to assure the members of the committee that the FDA had no intention of banning or significantly restricting the public's access to dietary supplements, as long as they presented no safety concerns. However, he did make a point of emphasizing that some dietary supplements do present safety concerns, and the FDA needed effective enforcement options for quickly dealing with these dangers. Second, he wanted to explain the FDA's reasoning for keeping the "significant scientific agreement" standard for health claims on dietary supplements. He felt strongly that health claims on dietary supplements should be held to the same standard as that for conventional foods. And Dr. Kessler referred to many of the concerns related to dietary supplements that we have also addressed in earlier chapters of this book. For example, the potential "indirect risks" associated with dietary supplement use (..."patients who forsake therapies that offer some real benefit for the siren song of empty promises..."). And the proliferation of "misleading claims and unfounded claims" that occurred before the passage of the NLEA in 1990. Other recurrent themes from earlier chapters will become apparent as the hearings proceed.

The next short excerpt is from an exchange between Senator Hatch and Mr. Michael Taylor, deputy commissioner of the FDA at the time. I include it here because it directly addresses the amino acid issue I discussed in a previous chapter.

Senator Hatch: *Senator Metzenbaum says that there is no concern about any desire to make vitamins and minerals, herbal products, or amino acids prescription drugs. Your own regulations on page 38 say that the task force recommended that amino acids containing dietary supplements be regulated as drugs. Am I wrong on that? Go ahead, Mike.*

There is a task force report which you have cited and which we have put out for public comment. But the agency has never proposed to regulate amino acids.

Mr. Mike Taylor (FDA):

Senator Hatch: *Not yet. Not yet you have not. But that is what is worrying people all over this country when you can say something like that. Now, it is not even regulation of drugs. Of course, I do not think anybody at the FDA would dare do that to vitamins and minerals.*

What the problem is an approval process that prices these products right out of the marketplace, and some of them, I suppose, might even be considered prescription drugs.

I think my time is up, but I will come back later in the next round.

You may recall from the previous chapter that the FDA had established a "task force" in 1991 to "review the agency's regulatory program for dietary supplements and to recommend improvements." One of the recommendations of this task force, in part as the result of the L-tryptophan incidents, was to regulate all amino acid-containing dietary supplements as drugs. Not surprisingly, this caught the attention of Senator Hatch who was clearly concerned that, if this task force recommendation eventually became an official regulation, it may lead to other dietary supplements being similarly regulated as drugs. He also noted that the enormous cost of getting a new drug approved through the FDA would "price these products right out of the marketplace."

The excerpt that follows includes an opening statement from Senator Tom Harkin (D-Iowa) as well as some of his questioning of Dr. Kessler and Mr. Taylor from the FDA. I have included it here because it illustrates some of the challenges associated with setting public policy on issues related to health, and the role that science (or sometimes pseudoscience) plays in this process.

Senator Tom Harkin

(D-Iowa)

Opening Statement

Thank you very much, Mr. Chairman. I am sorry I was late for opening statements. I just want to make a couple of comments before I get into questions.

Dr. Kessler, as you know, I have a long history of being interested in alternative medicine. I was the one, through my subcommittee on appropriations, that started the Office of Alternative Medicine at NIH 2 years ago because I have for a long time felt that we needed to take a look at alternative medical practices, therapies, and medicines; and also to break down the bias in medical research against the review of worthy treatments that are not in the mainstream of conventional medicine.

I also want to point out that our traditional health care system emphasizes high-technology medicine, and I think too often dismisses approaches that may be less costly and more preventative in nature. I just do not believe that conventional wisdom is always right and that mainstream medicine meets the needs or demands of everyone.

I have had a lot of publicity in the last few months. I have been suffering from allergies for years. Doctors prescribed Seldane; they prescribed everything for me. Finally, they said, "You have to start getting shots, Harkin." Until finally someone said this spring to me, "Have you ever tried bee pollen?" I said, "No, never heard of it." I started taking bee pollen. I have not had any allergies since. [Applause.]

They can clap some more. I do not know what is wrong with that. [Laughter.]
But I have been taking this bee pollen, and it has taken care of my allergies. And I do not take any other drugs. And no doctor ever prescribed this to me. And it does not say on it anywhere that it will cure my allergies. Obviously, the person that makes this said it would. My point is that nothing in this product is going to hurt me. I read all the ingredients. As a consumer, I wanted to know what was included and there is nothing that is going to hurt me in this product. It is a food supplement. So why shouldn't I take it and try it?

I think there are a lot of people around this country that are looking for other things to take other than drugs to try to cure some of their ailments. I point that out because I want to make sure that we are going down the right road. I want to make sure that people have access to these products. I also believe they should be informed and that is why I started the Office of Alternative Medicine because I want some of these things looked into, whether it is cancer therapies or a help with allergies. I want consumers to have a little bit more control over their own health care.

Consumers who take supplements have run into a bureaucracy that I believe has not been thoroughly objective and open to the growing body of evidence that indicates the values of dietary supplements and vitamins and other products. Consumers believe that the FDA wants to place unwarranted and arbitrary limits on vitamin and mineral dosage limits and regulate all amino acids and herbal products as drugs. Again, the FDA needs to clarify its position on this.

Last, let me just say this: We do need to strike a proper balance. We need to ensure access to safe products that show promising health benefits, but at the same time protect the public from harmful products and misleading claims. My experience with the Office of Alternative Medicine at NIH tells me that overcoming institutional bias is very tough. Very tough. So, again, we need to find a solution to this problem that will not leave the decisions about health claims entirely up to a bureaucracy that has time and time again shown an unwillingness to objectively weigh the evidence and apply the standards set forth in the Nutrition Labeling and Education Act in an appropriate manner.

Senator Harkin: *Having said all that, let me again say that I think that the NLEA, on which I worked for 10 years as a member of this committee and also the Agriculture Committee—is a good agreement. Are you telling me that what you want to see happen with vitamins and supplements would comport with what we have done in NLEA?*

Dr. Kessler: *That is correct, Senator.*

Senator Harkin: *One of the problems we had with NLEA and with the FDA is the issue of significant scientific agreement. Is this standard less than the standard for drugs. Everyone agreed on that. Now, maybe we did not do our job properly. We did not define what significant scientific agreement is. It has come to my attention from various sources that what you have applied thus far under NLEA for significant scientific agreement is not 51% or 55%, but more like 80 or 90%... Now what I would like to know is: What standard will you apply for significant scientific agreement for dietary supplements and vitamins?*

Mr. Taylor: *As Dr. Kessler said earlier, we recognize the law is very clear that the significant scientific agreement standard is a more flexible standard than the drug standard. It is more flexible in terms of the kinds of evidence we can consider. We are not required to have adequate and well-controlled clinical trials to reach this finding under NLEA. I think it is also more flexible in terms of the degree of certainty.*

When we approve a drug under the drug standard, we have a high degree of certainty about the efficacy of that product. Every...

Senator Harkin: *Excuse me. Thank you. And I do not mean to interrupt you. I take the admonition of the chairman seriously that we should not interrupt. I guess what I am trying to get at is: Do we need to spell out for you—and perhaps we should—what we mean by significant scientific agreement? Should it be 51%? I ask you that: Should it be 51%?*

Dr. Kessler: *Senator, I think it certainly would be worthwhile sitting down—and we are prepared to work with the committee—to come up with what was meant by significant. The problem ends up being 51%, you know, of what? In no two cases, 51% of the members of the National Academy of Sciences or the Alternative—it becomes hard to come up with a precise definition when you can plug it into a computer and an equation. That is the hard part.*

Senator Harkin: *That is right. And that brings me to the second part of the question. Does the agreement, whatever we would agree on, does the agreement have to only reflect studies published in major medical journals which often have a bias against accepting studies about nutrition in general? How are we going to set this up? One of the reasons I wanted to set up the Office of Alternative Medicine—and we set up a board, an advisory board—was to have lay people and medical people involved and who do not have an institutional bias against alternative medicines.*

I guess my question to you is: Do you see this as a possible way for the FDA to approach this kind of problem? Is there a role for that kind of advisory board made up of nontraditional medical researchers? Is there a role for nontraditional medical journals?

Dr. Kessler: *Senator, NLEA does require, as written in stat-ute, published studies. But on your point, I would welcome the Director of the Office of Alternative Medicine at the NIH to serve on an advisory com-mittee to be able to do these kinds of...*

Senator Harkin: *We can all clap for that.*

Dr. Kessler: *I have no problems with that. I think the people should be grounded in science. I think the data should be in public view, not in private view. I think the data should be open to everybody.*

Senator Harkin: *I also believe that you ought to have some lay people on that board, too, some people that are out there that maybe are not medical doctors but have valuable experience. I do not care whether it has been in homeopathic procedures or acupuncture or whatever it might be. But there are others out there that I think can bring a wealth of experience and knowledge to this kind of a process.*

Dr. Kessler: *There are consumer representatives on every advisory committee, and there is no reason why a consumer representative should not be on this advisory committee.*

Senator Harkin: *I cannot resist this, since Senator Pell brought up tree bark. I had an individual in my office a couple of weeks ago from New Mexico, and I told him I had a sore throat that day. He reached into his pocket, and he brought this out. He got it from some Native Americans in New Mexico. I don't remember what he called it. He said, "Break off a piece and chew it," and sure enough, it was the best anti-sore throat medicine I have ever used. [Laughter/applause.]*

So I am just telling you, there are things out there that people are using. Native Americans are using this treatment. I will break you off a piece if you ever have a sore throat. It will help you out. [Laughter.]

Senator Metzenbaum: Dr. Kessler, isn't it a fact that both that little piece and the pills that Senator Harkin is taking for—what is it?

Senator Harkin: My allergies.

Senator Metzenbaum: For his allergies. You do not have any problem with that as long as there is no misrepresentation about it?

Dr. Kessler: Absolutely. No problem.

Senator Harkin: But there should be some way to provide information—now, I do not say that this is going to cure everybody, just like Seldane quit working for me.

Dr. Kessler: It would be nice to get the data.

Senator Harkin: But it would be nice for the manufacturer—to be able to say that in certain cases, and in many instances, people who have taken this have been cured of allergies. What is wrong with that?

Dr. Kessler: Senator, the problem is what is the level of proof you want to establish and whether you want to just allow everybody to make any preliminary claim on a product. What is the marketplace going to look like? What are the aisles going to look like?

Senator Harkin: That is why I agree with you there should be significant scientific evidence. That is what we are trying to figure out here. If you are going to set the same standard as drugs, I am not in favor of that. If it is the same as NLEA, I think we can live with that, if, again, it is not as tight as what the drug is and if we have an advisory board or a group that can come up with this evidence that is not biased toward the traditional forms of medicine.

Dr. Kessler: *I do not disagree with that at all, Senator.*

Senator Harkin: *Thank you very much.*

Many of Senator Harkin's comments and questions reflect a common attitude among the general public regarding dietary supplements and the role that science should play in their regulation. Senator Harkin describes his anecdotal experience with bee pollen for his allergies, or tree bark for "whatever," or the "Native American" product for his sore throat. And he does not understand why manufacturers should not be allowed to promote these products, even if the data do not meet the "significant scientific agreement" standard. He is also clearly concerned about what he views as "institutional bias" against alternative treatments and alternative medicine. But Dr. Kessler is a scientist and he tactfully tries to respond to Senator Harkin's concerns with a reminder that scientific data in publically accessible, peer-reviewed scientific journals has to be the basis for distinguishing between legitimate versus unsubstantiated health claims. Some of these same issues and controversies can be noted in Senator Bingaman's questioning that follows.

Senator Jeff Bingaman
(D-New Mexico)

Thank you very much, Mr. Chairman.

Doctor, let me ask about some of your comments in your testimony, your written testimony, related to herbs. Some of them, you indicate that, for example, on—and I know nothing about these particular herbs. Let me just preface my statement that way, but it says "germander." I guess that is the way you pronounce it, "germander." You say there is a clear temporal relationship, these cases show a clear temporal relationship between ingestion of germander and onset of hepatitis.

Dr. Kessler: *Right.*

Senator Bingaman: *As you understand your authority right now, you do not have authority to do anything about the sale of unsafe herbs such as this that are on the market?*

Dr. Kessler:	*We can request a voluntary recall and have done that with a number of products.*
Senator Bingaman:	*But you have no authority to do anything more than request voluntary action?*
Mr. Zeller:	*That is right, without initiating case-by-case litigation. That is right. We cannot go in and wipe the shelves clean of...*
Senator Bingaman:	*No, I understand. But why can't you initiate case-by-case action against herbs that you believe are causing hepatitis?*
Mr. Zeller:	*Are you talking about litigation?*
Senator Bingaman:	*I am talking about any kind of action. If, in fact, you believe that there is a relationship between ingestion of this herb and the onset of hepatitis, why aren't you out there going to court or going somewhere to deal with this problem?*
Mr. Taylor:	*Under the laws that we work under, we can take enforcement action through the courts to try to remove those products from the market. The burden of proof that we bear in the case of a herb that is sold simply as a herb is that we must show that the product is ordinarily injurious to consumers, which means that there has to be a very high level of risk, a very high likelihood that people will be hurt.*
	That is one of the problems we have in this dietary supplement area, where the courts are saying your food additives safety standard does not apply. We are left with this far less effective standard, and that is why we say we do not want to have—we are comfortable bearing the burden to identify products that present demonstrable hazards. Let's just be sure we have an efficient tool for doing that, and we have got some real concerns about the adequacy of that tool.

Senator Bingaman:	*Where is that standard that applies in the case of herbs found?*
Mr. Taylor:	*It is in the food adulteration provision of our statute, Section 401(a)(1).*
Senator Bingaman:	*And you have to show that it ordinarily causes health hazards?*
Mr. Taylor:	*If it sold simply as a single-ingredient supplement, the courts are saying we have to show that it is ordinarily injurious.*
Senator Bingaman:	*And it is not the same standard that you have to show in the case of broccoli? If, in fact, there were a bunch of cases that showed there was a relationship between ingestion of broccoli and the onset of hepatitis, you would have an easier time going against the broccoli sales?*
Mr. Taylor:	*The law distinguishes between substances that naturally occur in food and that are sold as a single food ingredient, like the herb or like the broccoli. And in both cases, we have to meet this very serious burden of proving that the food is ordinarily injurious. If you add a substance to the broccoli, then it is a slightly easier standard for us to meet to prove harm. And if it falls within the legal definition of food additive, then we have the ability to shift the burden of proof to the company. So it is a very elaborate legal scheme, but in the herb case, the burden would be on us, your example, to prove the substance is ordinarily injurious.*
Senator Bingaman:	*Well, it just seems to me it is disturbing to see testimony saying that there is a causal tie—I guess that is what I understand—a clear temporal relationship. I assume that means if you eat the one you get the other.*
Dr. Kessler:	*One happens after a period of time. It is not a necessary cause and effect, but it certainly goes toward that cause and effect.*

Senator Bingaman:	*If, in fact, you said the same thing about broccoli that you are saying about germander, I would expect you to be out taking legal action to get broccoli off the shelves. If that meant going to court, that is what it would mean. But it just strikes me that there ought to be authority in the law—and I think there is today—for you to take action against unsafe herbs.*
Dr. Kessler:	*There is the authority, but under the "ordinarily render injurious" standard. The standard of "may render injurious," which we apply to any added substance in food, which is the possibility of risk, does not apply under recent court decisions to products such as single-ingredient herb products.*
Senator Bingaman:	*And have you asked us to change that law?*
Dr. Kessler:	*I think the whole safety question—I mean, that is not—I mean, the whole safety question needs to be thought about so that we can have some kind of safety review, I mean, that is thoughtful, that leads products on unless there is a demonstrable hazard. You know, a number of other countries have been much more aggressive. I know there are members who say that we have done too much in dietary supplements. You have every right to say we have not done enough. There are other countries that have acted against and banned whole classes of dietary supplements that we have not done in this country.*
Senator Bingaman:	*Yes. I guess the only points I would make are I think there is a clear difference between action that you should take and be able to take to protect the safety of the public—that is on one side, where I think you should have clear authority and there should be no question about it—and then the other issue of whether or not labels about potential benefits are misleading.*
Dr. Kessler:	*I agree with that, Senator.*

Senator Bingaman: *Substances which we all agree do not cause any harm, but may not cause the benefits, may not bring about the benefits that they are advertised to bring about. I think that is another issue. Now, on that, for example, there are a lot of folks in my State who drink chamomile tea on the theory that it helps them to sleep. In fact, you know, you can buy Sleepy Time Tea which Celestial Seasonings sells, and they say on the outside—or maybe they do not on that particular company's advertising, but the statement is generally made that this helpful for people sleeping. I do not know if there is any scientific basis for that. I would not be surprised to find out there is not. It does not seem to me particularly harmful, though, to be telling people that this traditionally has been thought to help people sleep. In my State, there are a lot of folks who sleep better at night thinking that it helps them. [Laughter.]*

Dr. Kessler: *Senator, the product should be available. There is no question about that. But if you allow products to be sold with claims, where is the incentive? You can just put claims on that do not have scientific evidence. Where is the incentive? Who is going to develop the evidence for where these products work? That is the issue. If anyone could go put any claim on the product, then the issue is how are you going to get the evidence on what works and what does not work. And in the end, we would all like to know, whether it is allergies or sore throats or treatment of insomnia, we would like to get the data. There is no incentive, Senator, if you can just put anything you want on the label.*

Senator Bingaman: *Well, maybe I am slicing it too thinly, but, I mean, the main point that I am getting at, I guess, is that some of these traditional herbs that have traditionally been thought to have certain benefits, I do not see that it hurts to be able to say these have traditionally been used with the understanding that they cause these benefits or with the expectation that they cause these benefits. Maybe that*

	is slicing it too thin. I do not know if they cause those benefits or not or bring about those benefits, but it would strike me that we are not doing any great harm to the folks in my State, as long as the herb is safe, allowing it to be sold as it always has.
Dr. Kessler:	*What about an herb for diabetes?*
Senator Bingaman:	*Well, I think that is a different business because there you are trying to cure a disease. In the case of suggesting that a herb helps with sleeping, that is not, in my view, trying to cure a disease. That is...*
Dr. Kessler:	*Senator, L-tryptophan was used for insomnia, and we saw real risks associated with that.*
Senator Bingaman:	*But that was not because it was mislabeled. That was because it was improperly manufactured. And I support your claim or your desire to ensure that manufacturing processes are appropriate.*
Dr. Kessler:	*Senator, we just need to be careful on the fact that L-tryptophan was due to manufacture. The science is still out on that. There is a substantial body of science that questions that hypothesis that it was due to a contaminant and not also due to the L-tryptophan itself. We are seeing that kind of disease associated with other brands and, in fact, with other amino acids. So I think we just need to be careful whether it is associated with the manufacturing or not.*
Senator Bingaman:	*My time has expired. I apologize to the Chairman for going on too long.*

Senator Bingaman's comments and questions continue along the theme started by Senator Harkin. That is, what is wrong with allowing companies to make health claims, even if the science is weak, as long as the risks are very low? Again, Dr. Kessler responds by warning that allowing some companies to make these claims without meeting the "significant scientific agreement" standard will open up the doors to many more potentially dangerous claims for other products. This exchange between Senator Bingaman and Dr. Kessler also focuses on the various "safety standards" that apply for different substances in the food supply. We will be exploring the issue of "safety standards" in much more detail in a later chapter of this book. But for now, simply note Dr. Kessler's reference to the "ordinarily render injurious to health" safety standard for some food substances versus the "may render injurious to health" safety standard for other substances. And Dr. Kessler alludes to the "food additive" regulatory authority that we discussed in the previous chapter.

The final exchange that I am including in this chapter is that between Senator Hatch and Dr. Kessler. It may be difficult to detect in this written transcript, but in the original video it is easy to sense the tension and clashing views of these two individuals.

Senator Hatch: *Now, Dr. Kessler, FDA's proposed regulations for health claims for dietary supplements set out four tests of preconditions that must be met before a dietary supplement manufacturer can petition the agency to approve a health claim. These preconditions are as follows:*

One, the dietary substance must be associated with a disease or health-related condition for which the general U.S. population is at risk, or the relevance of the claim must be explained within the context of the daily diet;

Two, the supplement must be a food. A food is a substance that must contribute taste, aroma, or nutritive value and retain that attribute when consumed at a level necessary to justify the claim;

Three, the substance must be safe and lawful under applicable U.S. food safety provisions of the Food, Drug, and Cosmetic Act;

Four, the health benefits must come from the nutritive value of the substance and not from the physiological process provided by the substance.

Now, please give me an example of any herbal dietary supplement that you believe could meet all four of those preconditions?
Are there any? I do not see any.

Mr. Taylor:	*Senator Hatch, these conditions that you have described are inherent in the current law as Congress has passed it, and the…*
Senator Hatch:	*My point is: Can any of them meet that?*
Mr. Taylor:	*That is a question that those who would want to submit claims to us under NLEA for an herbal product would address. Some no doubt…*
Senator Hatch:	*Well, you can see why they are concerned, can't you? If they cannot meet all four of those, they are dead, according to you.*
Mr. Taylor:	*Well, that is why…*
Senator Hatch:	*And there is no way they can meet…*
Dr. Kessler:	*But no one is talking about any of these products going off the market. The issue is whether they can make certain health claims and labels on the product, I mean, that are associated with the products. None of these products we are talking about, Senator, has to go off the market.*
Senator Hatch:	*That is precisely the issue. There is no question about it.*
Dr. Kessler:	*Right.*
Senator Hatch:	*And if these were pharmaceutical drugs, I can see your point. These are not. These are products that have been in existence for centuries that people have benefited from.*
	It is my understanding that the agency's policy is to send warning letters to prevent dietary supplement companies from providing information on their labels such as cautions, warnings, or specific dosage recommendations because such information makes these products new drugs. Is that correct? Mr. Taylor?
Mr. Taylor:	*Well, under the current laws that exist, a product is either a food or a drug. And if you make a disease-related claim for a product that does not fit within the food part of the statute, then under current law, the available remedy is the drug authorities. And we have used those authorities.*

But this is precisely why Congress is interested, and we agree that there ought to be an effort to recognize that dietary supplements have attributes that, as a practical matter, place them somewhere between what people think of as foods and what people think of as drugs. But under the current statute, we have those two choices to make.

Senator Hatch: *OK. Dr. Kessler, the agency recently gave 60 FDA employees awards for their role in attempting to remove evening primrose oil from sale. Now, what safety hazard was the FDA addressing that warranted such intensive use of agency resources and personnel?*

Dr. Kessler: *Senator, I can read you the claims made for oil of evening primrose. The list starts with cancer, Raynaud's syndrome. I mean, the list is about 20 or 30. Let me submit those for the record.*

Senator Hatch: *Remember, the issue is safety I am talking about.*

Dr. Kessler: *The claims...*

Senator Hatch: *Do you know of any unsafe...*

Dr. Kessler: *Gamma linolenic acid, and the courts that have looked at that have concluded that the agency's concern about safety was valid. My real concern, though, my real concern is the types of diseases for which oil of evening primrose is promoted, and I would be happy to submit that list for the record.*

Senator Hatch: *But my question is: What proof do you have that this substance is unsafe? I did not ask you what speculations you have. I asked what proof do you have. I mean, I had a Nobel Prize winner come in from Great Britain and tell me that this has been a very beneficial product.*

Dr. Kessler: *For what disease, Senator?*

Senator Hatch: *He could not even meet with you.*

Dr. Kessler: *Again, I mean, this is being promoted for a lot of different diseases, anywhere from hypertension to atopic dermatitis.*

Senator Hatch: *Safety, doctor, safety. That is the question. It is not?*

Dr. Kessler: *I would be happy to submit for the record the evidence that we submitted in court on animal studies that raised certain questions. But my major concerns about these products are the types of claims that are being made.*

Senator Hatch: *All right. Let me go to claims, but just one final question on safety. Is an American citizen more likely to die from an adverse reaction to a drug approved by the FDA or a dietary supplement?*

Dr. Kessler: *Senator, I am amazed. What do you think—what are in pharmaceuticals? I mean, half our pharmaceuticals come from natural—from plants.*

Senator Hatch: *What are in dietary supplements?*

Dr. Kessler: *Many come from plants, too.*

Senator Hatch: *Right.*

Dr. Kessler: *The origin—I mean, there are chemicals in pharmaceuticals, and those chemicals are found naturally. There are naturally occurring substances in dietary supplements. There is the assumption, you know, that all pharmaceuticals are toxic and natural substances are not, and I think that that belief—I mean, I just think we have been proven wrong on a number of occasions.*

Senator Hatch: *It sounds to me, though, like you are saying dietary supplements are the same, they are drugs. And, see, that is what worrisome to a lot of people in this industry, too.*

Well, let me go to claims because that is a very important part of this.

Dr. Kessler: *Senator, the issue...*

Senator Hatch: *That is what you are concerned about.*

Dr. Kessler: *The issue is, I mean, they are molecules. And you asked me about what kind of harm things can occur from dietary supplements. And there are instances of real harm. I agree with you.*

Senator Hatch: *I would like you to document them for me because I do not share that same overall, overriding concern that you do.*

Dr. Kessler: *The industry, Senator, agrees that there are risks with certain dietary supplements.*

Senator Hatch: *Sure, and they are very careful in the industry, by and large, to solve...*

Dr. Kessler: *And, Senator, I would appreciate—I mean, we have seen instances where the industry is not following its own guidelines on niacin, selling sustained release where the industry association is saying it should not be sold, selling Vitamin A in doses above what the industry sold, selling Vitamin B6 at above what the—I mean, I would be happy to submit that for the record.*

I am not saying—I do not want to exaggerate the safety concerns here. I said earlier I do not lose a lot of sleep. There are certain areas where I have certain concerns. We have some concerns about the amino acids, and I think we have to work it out.

Senator Hatch: *Let's work on it together and see if we can do something about it. I share your concerns about dietary supplement products that make claims that they can cure diseases without any or even sufficient scientific evidence or history to validate those claims. On the other hand, to give the other side of the coin, FDA has only approved a single health claim for a dietary supplement in 30 years, and that is, of course, calcium in osteoporosis in women, white and Asian women.*

Dr. Kessler: *Senator, the authority, as you said, was given to us to approve health claims for foods. It was given to us in 1990.*

Senator Hatch: *But our problem is that the agency also tries to prevent companies from making statements of general nutritional fact, and the agency apparently wants even to ban health food stores from distributing a variety of books, government documents, and even medical reports. Now, it is my understanding that, when promotional literature is* *making an unsubstantiated claim, the FDA believes that such literature containing the claim should be removed from the marketplace. It is also my understanding that the term "labeling" could include everything from pamphlets, books, brochures, to oral statements made by sales people. Am I incorrect on that?*

Dr. Kessler: *The definition of labeling is an expansive one, as upheld by the courts in the last 50 years of food and drug law.*

Senator Hatch: *Well, if that is so, do you believe that the book, "The Miracle Nutrients, Coenzyme Q10," which is listed in your report as a product making an unsubstantiated claim, should be removed from the marketplace?*

Dr. Kessler: *Senator, there is a spectrum. Senator Kassebaum and I talked about that spectrum of information. On the one hand, you have, you know, the* New England Journal. *On the other hand, you have promotional materials. I think that is something that we need to look at and talk about independent, third-party, peer review information, if it is not promotional in disguise. You and I see a lot of stuff that is presented, and it is made out to be independent, thoughtful evidence, thoughtful documentation, and it is nothing more than promotion in disguise. So it is a difficult question, and we need to be able to deal with that question.*

Senator Hatch: *All right. Could a dietary supplement product use literature which makes the following claim, "An increased intake of chromium could increase the glucose tolerance of many individuals, and thus might reduce the risk of heart disease"?*

Dr. Kessler: *Senator, I was told yesterday—I would be happy to submit it for the record. There are some safety concerns, as I understand it, with chromium that I would be happy to submit. I am not an expert on chromium.*

Senator Hatch: *I am talking about the claim. Can they make that claim? Would they be able to make that claim?*

Dr. Kessler: *If you could just restate it?*

Senator Hatch: *The actual quote that I gave you was, "An increased intake of chromium could increase the glucose tolerance of many individuals, and thus might reduce the risk of heart disease."*

Dr. Kessler: *That is a disease-related claim, on the surface of it.*

Senator Hatch: *They cannot make it in your eyes?*

Dr. Kessler: *That looks like a disease-related claim.*

Senator Hatch: *What about the following: "Persons with rheumatoid arthritis have a negative nitrogen and calcium balance. To control the progress of this disease, it is important to enhance protein and calcium ingestion"?*

Dr. Kessler: *Senator, I would have to look at that language. I could not comment on that.*

Senator Hatch: *Well, these two statements come from a Department of Agriculture report on human nutrition. Now, should that literature be banned?*

Dr. Kessler: *Senator, the Department of Agriculture is not selling dietary supplements, and I have no problems with independent parties making statements that are based on science. The issue is when the manufacturer uses statements to promote a product. That is where 50 years of food and drug law separates the third party, the independent statements based on science from the manufacturer using something to promote it.*

Senator Hatch: *Well, as you know, the Centers for Disease Control and Prevention has issued a recommendation for folic acid. Could a manufacturer of folic acid or health food store use this recommendation in conjunction with the sale of folic acid today?*

Mr. Taylor: *Again, where the law draws the line today—and I think we believe the law should draw the line—is when companies want to link a claim to a particular product and use it to promote and sell the product. It simply needs to meet the significant agreement standard.*

Senator Hatch: *But you did not answer the question, as far as I am concerned. I am saying, could they make that recommendation in conjunction with the sale of folic acid today? You are saying yes or no?*

Mr. Taylor: *I am saying...*

Senator Hatch: *You are saying they cannot, right?*

Mr. Taylor: *If they are using that claim to sell the product, the law today says—and NLEA stands for the principle that they have to have met the scientific standard.*

Dr. Kessler: *And we propose to approve that statement.*

Senator Hatch: *Six months from now, if we are lucky. In other words, the point I am making is that the poor little health food storeowner could not even hand out a government pamphlet from Centers for Disease Control or from the Agriculture Department.*

Dr. Kessler: *Senator, we have not said that. We have not said that. You asked me whether—you asked me if it is used to promote a specific product, if it is used to accompany a product. If there is independent literature and it is not associated with individual promotion, and it is really true independent literature, I think that is something—there is a spectrum, and I think that is something that we need to sit and consider.*

Senator Hatch: *In your report on unsubstantiated claims by store employees, several of those employees first consulted a book titled* Prescription for Nutritional Healing *by James F. Bolch, M.D.—I do not know if I am pronouncing that name right—and Phyllis A. Bolch, C.N.C. The book is based upon their experience using dietary supplements in patient care. Does the FDA believe that this book should not be available as a reference tool for employees or customers of health food stores?*

Dr. Kessler: *The FDA, what we tried to do in that list was to tell you exactly what our experience was. We are not saying one way or the other whether that is appropriate. We have not taken enforcement actions on those particular areas.*

Senator Hatch: *That still does not answer the question. Can they use that book? Can they refer to it?*

Dr. Kessler: *I would be happy to study that book, Senator.*

Senator Hatch: *All right. I would like you to do that. I think you might add to your store of medical knowledge if you would. [Laughter.]*

Chairman Kennedy: *We will have order in the audience now. These witnesses are responding to various questions, and we will ask that the audience be courteous in their response and not demonstrate either approval or disapproval. That is the way this institution has worked and will continue to work.*

Senator Hatch: *Thank you, Mr. Chairman.*

Let me give you a hypothetical that is important. A customer walks into a store and says that he has heard in the news that he should take Vitamin E. The employee quotes a recent article in CSPI's October health letter, which suggests that while researchers will not know for sure for several years whether antioxidants can help prevent heart disease, it makes sense to take antioxidants like Vitamin E, beta carotene, and Vitamin C every day. On the basis of that testimonial, the customer buys products which provide the dosage as recommended by CSPI. Would the employee's use of CSPI's newsletter constitute an "unsubstantiated claim"?

Dr. Kessler: *I would be happy to provide you with an analysis of that.*

Senator Hatch: *But what is your feeling?*

Dr. Kessler: *Senator, I am trained as a lawyer, and you know I am not going to give you a legal opinion. First of all, I am not going to give you a legal opinion anyway because I...*

Senator Hatch: *I want you to put your legal head aside and tell me as the head of the FDA if you think that is an unsubstantiated claim.*

Dr. Kessler: *I think that the oral representations—I mean, I happen to*
agree with Senator Kassebaum. I am less concerned about,
you know, the stores than I am about the information com-
ing from manufacturers. And I think that oral discussions
of what is in the New England Journal—I mean, I think
that we would like to get it right. I think that CSPI may be
wrong on the Vitamin C in that instance. The New England
Journal study showed no effect of that. It did show an effect
of Vitamin E. But I do not have a lot of—I mean, I think
oral representations that are done in good faith, that try to
capture stuff in the New England Journal, we are not going
to go after that, Senator. I have not gone after...

Senator Hatch: *I understand. Just two last thoughts, because we have*
kept you a long time and I have appreciated your patience
and the patience of my colleagues. In your testimony, you
refer to the importance of allowing consumers to make
informed choices about dietary supplements. But the only
information that you would permit these consumers to
have is that white and Asian women might take calcium
for osteoporosis. While Harvard Hospital releases a study
showing that Vitamin E may help prevent heart disease, a
manufacturer or retailer could very easily violate the law.
As you are interpreting it, for telling its customers about
the study. And that does not make any sense to me, and
I am sure it does not to you if you really think about it.

One final observation. Consumers need information
to make informed choices, but the current regulatory
arrangement impedes instead of educates. It seems absurd
to me that Americans have to sneak copies of Government
reports, medical journals, scientific treatises to educate
themselves on how to lead healthier lives and help pro-
tect themselves and their families from spiraling medical
costs. It is time for the FDA to work with Congress to
develop a more intelligent approach. I would like to do
that. And let me just say this to you: There is nobody that
would exceed me in wanting to keep false and bad prod-
ucts off the marketplace. You know that.
I know that.

Dr. Kessler: *Absolutely.*

Senator Hatch: *But, you know, Senator Pell in his comments, if it does not hurt them, why are you giving them such a rough time about it? The fact of the matter is that many people get well because they take placebos, because in a large sense they believe they are doing something that helps them, and psychosomatically it does. A lot of doctors feel that 80% of all illness is psychosomatic, or at least psychosomatic-related. This is an industry where there is a very low incidence of risk.*

I do admire you and appreciate your efforts in trying to make sure the American consumers are protected. But it is an industry where you really cannot show much in the way of risk from a percentage standpoint, a statistical standpoint, or even an actuality standpoint. And the few times that you do, there are good answers for it. Very good answers. And this is not the pharmaceutical industry. This is not the chemical industry. And I think there has got to be a more open mind toward these.

Now, I would like to have you examine the products you brought here today, and really, if you do not mind, I would like you to leave those with us so that we can review them. I would like to examine them and just see what we think about them, if you do not mind. We will take a good look at them as well. And when there are false and misleading things, you have the total authority right now to take them off the marketplace. Where things are unsanitary or toxic or poisonous or deleterious, you have the total power right now to take them off the marketplace. But what a lot of people out there feel you have been advocating for, especially if you listened to your testimony before the House committees, is basically that they have got to prove every claim that they make before they can put the claims out there. And if they do not, they cannot do it. Therefore, a lot of products that basically are helping people like Senator Harkin—I remember when he started to take bee pollen. You know, I am a believer in bee pollen, but the number of pills that they were telling him to take every day was kind of exciting to me. I thought it was really something. But he did. And I knew he suffered tremendously from that, and he just got better.

Now, you know, when there is not much risk, I think there ought to be a little more leeway. And, frankly, what the FDA has been arguing for over the last number of years has been a lot less leeway. And I do not think your record is a good record with regard to approving claims, and I can see why nobody in this industry would want to leave it up to the FDA to approve claims when you have only approved 1 in 30 years and then one that is so clear-cut it is not even funny, and only 6 out of 11 approved of folic acid supplementation or fortification. You know, that has to bother anybody.

So these are some of the things that are bothering me. Now, I do want to sit down and work with you, with Mr. Taylor and you, and try to resolve this problem. I do not want bad products out there any more than the industry does. The industry has been tainted because of some of the accusations, frankly because of some of these displays that we have had at some of these hearings. And this is a good industry that does an awful lot of good for people, and there are millions, 100 million people who take these substances that feel that they are healthier and better. And I know doctors here in this room right now who are helping patients with AIDS with nutritional therapy to a much more beneficial effect than some are with the known pharmaceutical therapies. And that is not knocking the pharmaceutical therapies. I am just saying nutritional therapy can help in a wide variety of ways. I am sure of it. And I think others would back that up in the scientific community as well.

Well, I thank my...

Dr. Kessler: *Senator, we would be happy to make copies of the labels and give you those.*

Senator Hatch: *You are afraid we will use those? [Laughter.]*

Dr. Kessler: *No, Senator. I think that we stand ready to work with you. Our goals are the same.*

Senator Hatch: *I hope they are. Thanks, Mr. Chairman.*

Senator Hatch is clearly much more confrontational than his fellow committee members when questioning Dr. Kessler. He even resorts to attempt to trick Dr. Kessler into disavowing statements made by the United States Department of Agriculture (USDA) regarding chromium in the diet. And at one point, when Dr. Kessler states that he would be happy to review a book on nutritional healing that Senator Hatch mentioned, the senator sarcastically responds, "I think you might add to your store of medical knowledge if you would." In general, this is the Senator's style. But aside from his sarcasm and confrontational style, he brings up several points that are central to the controversies surrounding the regulation of dietary supplements. First, as with Senators Harkin and Bingaman, he challenges Dr. Kessler to defend the need for more strict regulation of dietary supplements due to safety concerns. Although Dr. Kessler acknowledges that the vast majority of dietary supplements are safe, he notes that some are not. In addition, Dr. Kessler attempts to get the Senator to focus on the issue of unsubstantiated health claims and the problems that they pose to the industry and to the ability of consumers to make informed purchasing decisions.

The other important issue raised in this exchange relates to the issue of "labeling" and the inability of dietary supplement manufacturers to include so-called "third-party literature" as part of the marketing of their products. According to the Food, Drug, and Cosmetic Act in effect at the time of this hearing, this third-party literature could be considered part of the "labeling" of the product. If the literature made disease claims related to the product, the product could meet the legal definition of a drug, as we discussed in Chapter six. This is worth keeping in mind as we move into the next several chapters related to the Dietary Supplement Health and Education Act of 1994.

chapter ten

DSHEA
Defining dietary supplements

The bill sponsored by Senator Hatch, S. 784, did not get voted on in the 1993 Congress. However, Senator Hatch introduced an amended version of the bill in the 1994 session of Congress. This amended bill included some minor concessions to the Food and Drug Administration (FDA). It was also intended to appease some members of Congress who felt that the original bill went too far in limiting the FDA's authority over dietary supplement regulation. The amended bill was passed unanimously in both the House and the Senate. On October 25, 1994, President Clinton signed into law the Dietary Supplement Health and Education Act (DSHEA). President's Clinton's signing statement clearly conveys the attitude of the government (although certainly not of the FDA) and the mood of the general public (although certainly not of the established scientific and public health community) regarding regulation and access to dietary supplements.

President William J. Clinton
Statement on Signing the DSHEA of 1994
October 25, 1994

Today I am pleased to sign S. 784, the "Dietary Supplement Health and Education Act of 1994." After several years of intense efforts, manufacturers, experts in nutrition, and legislators, acting in a conscientious alliance with consumers at the grassroots level, have moved successfully to bring common sense to the treatment of dietary supplements under regulation and law.

More often than not, the Government has been their ally. And the private market has responded to this development with the manufacture of an increasing variety of safe supplements.

But in recent years, the regulatory scheme designed to promote the interests of consumers and a healthful supply of good food has been used instead to complicate choices consumers have made to advance their nutritional and dietary goals. With perhaps the best of intentions agencies of government charged with protecting the food supply, the rights of consumers have paradoxically limited the information to make healthful choices in an area that means a great deal to over 100 million people.

And so, an historic agreement was finally reached in the Congress this year that balances their interests with the Nation's continued interest in guaranteeing the quality and safety of foods and products available to consumers. This agreement was embodied in S. 784, legislation sponsored in the Senate by Senator Orrin Hatch and Senator Tom Harkin, in the House by Congressman Bill Richardson, and passed with the help of Senator Edward Kennedy, Congressman John Dingell, Congressman Henry Waxman, and scores of cosponsors in the House and Senate.

Simply said, the legislation amends the Federal Food, Drug, and Cosmetic Act to establish new standards for the regulation of dietary supplements including vitamins, minerals, and herbal remedies.

The passage of this legislation speaks to the determination of the legislators involved, and I appreciate their work. But most important, it speaks to the diligence with which an unofficial army of nutritionally conscious people worked democratically to change the laws in an area deeply important to them. In an era of greater consciousness among people about the impact of what they eat on how they live, indeed, how long they live, it is appropriate that we have finally reformed the way Government treats consumers, and these supplements in a way that encourages good health.

WILLIAM J. CLINTON
The White House, October 25, 1994.

Throughout the long history of the FDA, most laws that the agency was charged to enforce were designed to strengthen or expand the FDA's power and authority to better protect the safety and interests of the consumer. However, not unlike the Vitamin and Mineral Amendment (Proxmire Amendment) of 1976, the DSHEA was essentially a law intended to limit or restrict the authority of the FDA to regulate a consumer product. Even President Clinton's signing statement delivers a gentle but clear message to the FDA, suggesting that the agency has been overly zealous in its regulatory approach to the dietary supplement industry. As President Clinton states

> But in recent years, the regulatory scheme designed to promote the interests of consumers and a healthful supply of good food has been used instead to complicate choices consumers have made to advance their nutritional and dietary goals. With perhaps the best of intentions agencies of government charged with protecting the food supply and the rights of consumers have paradoxically limited the information to make healthful choices in an area that means a great deal to over 100 million people.

The DSHEA has been the "law of the land" with regard to dietary supplement regulation for nearly a quarter of a century now. How successful it has been is still the subject of much debate. In this and subsequent chapters, I will review the major provisions of this law and consider them in the context of the FDA's regulation of other food substances.

The legal definition of a dietary supplement

The first and perhaps most significant provision of the DSHEA was the establishment of an explicit legal definition for the term, "dietary supplements." This section of the law (21 U.S. code (USC) § 321(ff)) has a number of important subsections, which I will discuss separately. The first section describes what types of substances would meet the legal definition of a dietary supplement.

21 USC § 321(ff)

The term "dietary supplement"—
(1) means a product (other than tobacco) intended to supplement the diet that bears or contains one or more of the following dietary ingredients:
    (A) a vitamin;
    (B) a mineral;

(C) an herb or other botanical;
(D) an amino acid;
(E) a dietary substance for use by man to supplement the diet by increasing the total dietary intake; or
(F) a concentrate, metabolite, constituent, extract, or combination of any ingredient described in clause (A), (B), (C), (D), or (E);

The first thing to note about this section of the law is that "intended use" is part of the determination of whether or not a substance can be classified as a dietary supplement. If it is *intended to supplement the diet*, it may be a dietary supplement. This section goes on to list all of the possible substances that may be used to supplement the diet. It is hard to imagine a more broad legal definition for these substances. Note that amino acids are explicitly included in this list, perhaps as a direct rebuke to the FDA's earlier suggestion that amino acid supplements may be regulated as drugs. And, less anyone feel that a possible dietary supplement substance was excluded in clauses (A)–(E), the law includes clause (F), which covers just about any other possible material that one might want to include within the legal definition of a dietary supplement.

The next subsection of this portion of the law goes on to further specify the factors to be considered when determining whether a substance meets the legal definition of a dietary supplement.

(ff) The term "dietary supplement"—
(2) means a product that—
(A)(i) is intended for ingestion in a form described in section 350(c)(1)(B)(i) of this title; or (ii) complies with section 350(c)(1)(B)(ii) of this title;
(B) is not represented for use as a conventional food or as a sole item of a meal or the diet; and
(C) is labeled as a dietary supplement;

The reference to Section 350(c)(1)(B)(i) in the above USC excerpt refers to the form of the substance that would meet the definition. Specifically, it states that in order for a substance to be considered a dietary supplement, it would need to be ingested in "tablet, capsule, powder, softgel, gelcap, or liquid form." Thus, items consumed in the form of a conventional food (or as the sole item of a meal or diet) would not meet the legal definition of a dietary supplement. This is a very important provision and one that we will discuss further in a later chapter. Finally, in order for a substance to meet the legal definition of a dietary supplement, it must be labeled as a "dietary supplement."

As we have discussed extensively in earlier chapters of this book, one of the most controversial issues related to the broad topic of dietary supplement regulation has been the debate regarding when the Food, Drug, and Cosmetic Act would consider a substance a "food" and when it would consider it a "drug." This is partially addressed in the next subsection of this portion of the law.

> 21 USC §321(ff)(3)(B) [The term "dietary supplement" does] not include—
>
> (i) an article that is approved as a new drug under section 355 of this title, certified as an antibiotic under section 357 of this title, or licensed as a biologic under section 262 of title 42, or
>
> (ii) an article authorized for investigation as a new drug, antibiotic, or biological for which substantial clinical investigations have been instituted and for which the existence of such investigations has been made public,
>
> - which was not before such approval, certification, licensing, or authorization marketed as a dietary supplement or as a food unless the Secretary, in the Secretary's discretion, has issued a regulation, after notice and comment, finding that the article would be lawful under this chapter.

This section may be a bit confusing, partly due to some "double negatives" in the phrasing. But it is a very important clause. What it essentially states is that "new drugs," or substances that are in the process of being studied as possible new drugs ("authorized for investigation"), do not meet the legal definition of a "dietary supplement" unless the substance that makes up the "new drug" was previously marketed as a dietary supplement. In other words, for example, if I am the manufacturer of a dietary supplement and sometime later a company develops a new drug based on the ingredient(s) in the dietary supplement, my original product could still meet the legal definition of a dietary supplement, even though it is also now a "drug." On the other hand, if the drug company develops their drug first, and then I try to manufacture a product based on the active ingredient in the drug, my product would not meet the legal definition of a "dietary supplement," since it was already a new drug (or an investigational new drug).

The section 321(ff) dietary supplement definition concludes with one more small but important clarification:

> Except for purposes of paragraph (g) and section 350(f) of this title, a dietary supplement shall be deemed to be a food within the meaning of this chapter.

This wording in the DSHEA was intended to remove any confusion regarding where dietary supplements fall within the regulatory scheme of foods versus drugs.

The DSHEA also modified the drug definition of the law in 21 USC §321(g)(1). The relevant portion of this definition included as part of the DSHEA is italicized in the USC excerpt later.

> 21 USC §321(g)(1) The term **"drug"** means
>
> (A) articles recognized in the official United States Pharmacopoeia, official Homoeopathic Pharmacopoeia of the United States, or official National Formulary, or any supplement to any of them; and
>
> (B) articles intended for use in the diagnosis, cure, mitigation, treatment, or prevention of disease in man or other animals; and
>
> (C) articles (other than food) intended to affect the structure or any function of the body of man or other animals; and
>
> (D) articles intended for use as a component of any article specified in clause (A), (B), or (C).
>
> *A food or dietary supplement for which a claim, subject to sections 343(r)(1)(B) and 343(r)(3) of this title or sections 343(r)(1)(B) and 343(r)(5)(D) of this title, is made in accordance with the requirements of section 343(r) of this title is not a drug solely because the label or the labeling contains such a claim. A food, dietary ingredient, or dietary supplement for which a truthful and not misleading statement is made in accordance with section 343(r)(6) of this title is not a drug under clause (C) solely because the label or the labeling contains such a statement.*

Reference to Section 343(r) in the earlier drug definition refers to various portions of the law that deal with health claims on food labels. So the DSHEA modified the drug definition to make it clear that a dietary supplement cannot be classified as a drug if its labeling bears a health claim that is in compliance with the law as specified in Section 343(r).

As you may recall from a previous chapter, the FDA had considered using its food additive regulatory authority to regulate ingredients in dietary supplements. The dietary supplement industry and its advocates in Congress vigorously opposed this strategy. As a result, the DSHEA included a small change to 21 USC §321(s). The relevant portion is italicized in the box below.

---

21 USC §321(s) The term "food additive" means any substance, the intended use of, which results or may reasonably be expected to result, directly or indirectly, in its becoming a component or otherwise affecting the characteristics of any food (including any substance intended for use in producing, manufacturing, packing, processing, preparing, treating, packaging, transporting, or holding food; and including any source of radiation intended for any such use), if such substance is not generally recognized, among experts qualified by scientific training and experience to evaluate its safety, as having been adequately shown through scientific procedures (or, in the case as a substance used in food before January 1, 1958, through either scientific procedures or experience based on common use in food) to be safe under the conditions of its intended use; *except that such term does not include—*

(1) a pesticide chemical residue in or on a raw agricultural commodity or processed food; or
(2) a pesticide chemical; or
(3) a color additive; or
(4) any substance used in accordance with a sanction or approval granted before September 6, 1958, pursuant to this chapter, the Poultry Products Inspection Act (21 USC 451 et seq.) or the Meat Inspection Act of March 4, 1907, as amended and extended (21 USC 601 et seq.);
(5) a new animal drug; or
(6) *an ingredient described in paragraph (ff) in, or intended for use in, a dietary supplement.*

By adding item (6) to the list of substances that are not considered food additives, Congress made it very clear that the FDA cannot use its food additive authority to regulate substances that meet the legal definition of dietary supplements. Of course, this only pertains to those substances in the product that meet the definition of a "dietary supplement" (according to 21 USC §321(ff)). Other "non-dietary supplement" ingredients in the product may still be regulated as "food additives." This may include flavors, binders, coatings, preservatives, etc.

For some ingredients that meet the definition of a dietary supplement, the DSHEA requires that FDA be notified before its use in a dietary supplement product. This applies to any so-called "New Dietary Ingredient." The DSHEA defines a New Dietary Ingredient as follows:

21 USC §350(d)
"New dietary ingredient" defined
For purposes of this section, the term "new dietary ingredient" means a dietary ingredient that was not marketed in the United States before October 15, 1994 and does not include any dietary ingredient which was marketed in the United States before October 15, 1994.

So any dietary ingredient that was part of the U.S. food supply before enactment of the DSHEA (on October 15, 1994) can be used in any new or future dietary supplement product without having to notify the FDA. The new dietary ingredient (NDI) provisions of the law (and the FDA's enforcement of these provisions) is another example of the sometimes murky distinction between the regulation of a "new dietary ingredient" in a dietary supplement, and a "food additive" in a dietary supplement. A dietary supplement NDI must meet the legal definition of a "dietary ingredient" in a dietary supplement. In other words, it must be "a vitamin; a mineral; a herb or other botanical; an amino acid; a dietary substance for use by man to supplement the diet by increasing total dietary intake; or a concentrate, metabolite, constituent, extract, or combination of any of the above dietary ingredients." (21 USC §321(ff)(1)). If it is a new ingredient that does not meet this definition, then the FDA may have the authority to regulate it as a "food additive." However, if the ingredient meets the definition of a "new dietary ingredient," then the following section of the DSHEA applies.

21 USC §350(b) New dietary ingredients

(a) In general
A dietary supplement which contains a new dietary ingredient shall be deemed adulterated under section 342(f) of this title unless it meets one of the following requirements:

(1) The dietary supplement contains only dietary ingredients that have been present in the food supply as an article used for food in a form in which the food has not been chemically altered.

(2) There is a history of use or other evidence of safety establishing that the dietary ingredient when used under the conditions recommended or suggested in the labeling of the dietary supplement will reasonably be expected to be safe and, at least 75 days before being introduced or delivered for introduction into interstate commerce, the manufacturer or distributor of the dietary ingredient or dietary supplement provides the Secretary with information, including any citation to published articles, which is the basis on which the manufacturer or distributor has concluded that a dietary supplement containing such dietary ingredient will reasonably be expected to be safe.

Thus, unless the ingredient was already present in the food supply, the manufacturer would be required to provide the FDA with evidence of its "safety" at least 75 days before the company intends to introduce the product into interstate commerce. I put the term "safety" in quotes here because the standard of safety for dietary supplements is unique to this category of food substances. This will be further explained and discussed in a later chapter. However, notice in the earlier section of the law that the DSHEA does not require that the manufacturer conduct any detailed safety or toxicity testing of the "new dietary ingredient." Rather, it is only necessary that it provide the FDA with "evidence of safety." The FDA has published regulations that provide a bit more detail on what types of "evidence of safety" it would require. In Title 21 of the Code of Federal Regulations (CFR):

21 CFR §190.6 Subpart B—New Dietary Ingredient Notification Requirement for premarket notification.

(b) The notification required by paragraph (a) of this section shall include:

(4) The history of use or other evidence of safety establishing that the dietary ingredient, when used under the conditions recommended or suggested in the labeling of the dietary supplement, will reasonably be expected to be safe, including any citation to published articles or other evidence that is the basis on which the distributor or manufacturer of the dietary supplement that contains the new dietary ingredient has concluded that the new dietary supplement will reasonably be expected

to be safe. Any reference to published information offered in support of the notification shall be accompanied by reprints or photostatic copies of such references. If any part of the material submitted is in a foreign language, it shall be accompanied by an accurate and complete English translation;

Both the law (the DSHEA) and the associated FDA regulations (in the CFR) only require that the manufacturer provide "history of use" and/or other "citations to published articles or other evidence" of safety. There is no requirement for extensive animal or human clinical studies to establish safety, as is the case for new food additives or new drugs. In addition, this notification process should not be construed as an "approval" by the FDA of the safety of the ingredient. It is simply a notification to the FDA that the manufacturer intends to use the new dietary ingredient and has provided some level of evidence of its safety. If the FDA has concerns regarding the safety of the ingredient (at the time of notification or any other time in the future), the burden of proof for establishing the lack of "safety" would lie with the FDA, not the manufacturer.

chapter eleven

DSHEA

A new safety standard for dietary supplements

One of the most controversial provisions of the Dietary Supplement Health and Education Act (DSHEA) has to do with the standard it established for determining whether a dietary supplement is "safe" within the legal meaning of the law. To fully appreciate why this is so controversial, it is necessary to first review some basic concepts of toxicology, in particular, the concepts of *safety* versus *risk*. We can then examine and compare how the DSHEA deals with determining the safety and risk for dietary supplements versus how the Food, Drug, and Cosmetic (FD&C) Act deals with the safety and risk of other substances in our food supply.

Let's start with one of the most basic principles of toxicology. Consider the following quote from the 15th century Swiss-German physician and philosopher, Paracelsus, often referred to as the "father of modern toxicology."

> *All things are poisons, for there is nothing without poisonous qualities. It is only the dose which makes a thing poison.*
>
> Paracelsus (1493–1541)

Nearly 500 years ago, he recognized and articulated this most fundamental principle of modern toxicology that every substance is potentially toxic. It is simply a function of dose. Even water could be toxic if it is consumed at high-enough levels (dose). Once you accept this scientific reality, it is then easy to recognize that there is no such thing as absolute (or 100%) "safety" of any food substance. There will always be some risk associated with ingestion of anything. And that risk will increase as you increase your exposure to the substance. So how do we define the term "risk"?

Toxicologists define risk as the *probability* that harm will result from exposure to the substance. The particular "harm" associated with the risk will vary depending on the particular substance. Consider the example of alcohol. Alcohol has many potential harmful effects, from disorientation, nausea, and headache, to more serious effects such as respiratory failure, liver disease, cancer, and death. But the important thing to note about each of these potential harmful effects is that the risk of experiencing them can be scientifically measured or estimated.

For example, scientists can expose animals to increasing doses of alcohol and measure how many of the animals exhibit a particular harmful effect at each test dose. At the low doses of exposure, few if any of the animals may exhibit the toxic effect. As the dose increases, many animals exhibit toxicity. At some higher level of exposure, all of the animals will exhibit the toxic effect. This is known as the "dose–response relationship," and it is a fundamental concept of toxicology and an essential consideration in the regulation of substances in our food supply.

The term "safe" or "safety" is much more difficult to define. There are a number of reasons for this. One of the most important is that unlike risk, which can be measured experimentally, there is no way to directly measure the safety of a substance. Toxicologists sometimes define "safe" as the *probability that no harm* will result from the substance. But, of course, it is not possible to measure something that doesn't happen. So instead, safety is sometimes expressed as the *reciprocal* of risk. For example, if we measure the risk of exposure to a substance at a particular dose and find that the probability that the dose will cause harm is 10% (10% of the animals exposed at this dose exhibit the toxic effect), then the probability of no harm occurring at this dose would be 90%. In other words, the "safety" of the substance at this dose is 90%. But the term "safety" in the real world (such as in the regulation of food substances) is much more complicated than a simple statistical probability estimate. Many other factors are typically considered when determining whether a food substance is "safe." Furthermore, which particular factors that are to be considered (and the extent to which they are considered) will vary depending on the nature of the particular food substance. This brings us to the concept of "acceptable risk." In many ways, the "safety" of substances in the food supply can be expressed as follows:

Safe = Acceptable Risk

We deal with the concept of "acceptable risk" every day. Consider again the example of alcohol. Society may consider it "safe" to drink alcohol in moderation. But, of course, it is not 100% safe. Even in moderation, some individuals will be more sensitive to the effects of alcohol and may be more likely to experience harmful effects (nausea, liver disease, etc.). But we consider it "safe" because as a society we consider drinking in moderation to be an "acceptable risk." Now consider the concept of acceptable risk in the context of food substances. The Food and Drug Administration (FDA) may have very different standards for acceptable risk (safety) for different food substances. For example, the acceptable risk for a food additive artificial sweetener may be very different from the acceptable risk for a food preservative that prevents growth of bacterial pathogens in the food. Or the acceptable risk for a substance that cannot be avoided in the food supply (such as a naturally occurring toxin in foods) may be very different from the acceptable risk for a substance that is intentionally added to a food. Congress recognized this when it wrote the FD&C Act and many of its amendments over the years. And the FDA also recognized this when it wrote the many regulations that enforce the "safety" provisions of the FD&C Act.

So let's now look at how the Congress and the FDA addressed the issue of *standards of safety* for different substances in the food supply and compare these to how the DSHEA deals with the safety of dietary supplements. First, it is important to understand that there is a "standard of safety" for everything in the food supply, whether it be a food additive, a pesticide residue, an environmental contaminant in the food, a piece of potato in a can of beef stew (a "generally recognized as safe" or "GRAS" substance)... everything. In the table below, I have listed nine categories of food substances, an example for each, and a brief description of the safety standard for each category. I adapted this table from an article published by Dr. David Kessler in 1984, well before he became Commissioner of the FDA.[1]

Food substance category	Examples	Safety standard
Food additives	Xanthum gum	Delaney
Color additives	FD&C yellow #6	anticancer
Animal drug residues	Bovine growth hormone	clause
		or
		"Reasonable certainty of no harm."
		(Continued)

[1] Kessler, DA. Food Safety: Revising the Statute. *Science*, 223: 1034–1040 (1984).

Food substance category	Examples	Safety standard
GRAS substances	Salt, sugar	The substance is safe if it is generally recognized, among experts qualified by scientific training to evaluate its safety, as having been shown through scientific procedures... or through experience based on common use in food, to be safe under the conditions of intended use.
Pesticide chemical or residues	Aldrin, dichlorodiphenyltrichloroethane (DDT)	A tolerance may be set that takes into account, among other things, the production of an adequate, wholesome, and economical food supply.
Prior-sanctioned substances	Butylated hydroxytoluene (BHT) (antioxidant)	The food is adulterated if the prior-sanctioned substance may render the food injurious to health.

(Continued)

Food substance category	Examples	Safety standard
Added poisonous or deleterious substances	Aflatoxin in peanuts	The food is adulterated if the poisonous or deleterious substance *may render the food injurious to health.*
Naturally occurring (not added) poisonous or deleterious substances	Solanine in potatoes	The food is adulterated if the poisonous or deleterious substance *renders the food ordinarily injurious to health.*
Added poisonous or deleterious substances that are required in the production of food or cannot be avoided by good manufacturing practices.	Polychlorinated biphenyls (PCBs) in fish	A tolerance may be set that takes into account the extent to which the use of such substance is required or cannot be avoided, at levels necessary to protect the public health.

There are nine different categories of food substances included in this table. Notice that there is no listing for "dietary supplements." That is because this list represents the categories of food substances that existed before the passage of the DSHEA. Recall from our discussion in Chapter eight that dietary supplement ingredients were generally classified as GRAS substances before the DSHEA was enacted. But recall also that the FDA's June 1993 Federal Register *Advanced Notice of Proposed Rulemaking* (ANPRM) suggested that it was considering regulating some dietary supplement ingredients as food additives.

Before the passage of the DSHEA, everything that we consume as food would fit into one of the nine legal categories of food substances in the above table. Not surprisingly, the vast majority of these would be those that are regulated as GRAS substances. To understand how and why various safety standards are used for these categories of food substances, I am going to focus on just three of the nine categories: food additives, added poisonous or deleterious substances, and naturally occurring (not added) poisonous or deleterious substances. I chose these three categories because they represent a nice range of safety standards, and because they represent safety standards that have some relevance to the topic of dietary supplement regulation. In the case of food additives, as just mentioned, the FDA had considered regulating some dietary supplement ingredients in this category. In the case of added and naturally occurring poisonous or deleterious substances, the safety standards that apply to these categories were the subject of some discussion in the Senate hearings related to S. 784 as presented in Chapter nine of this book.

Let us start with food additives. The first thing to notice from the table is that the safety standards for food additives are essentially the same as those for color additives and animal drug residues. For this discussion, I am just going to focus on food additives. The next thing to notice is that in determining if a food additive is "safe" (that is, presents an acceptable risk within the meaning of the law), two safety standards need to be satisfied. The first is the so-called *Delaney Clause*, added as part of the Food Additive Amendment to the FD&C Act in 1958. The Delaney Clause states the following:

21 U.S. code (USC) §348(c)(3)(A): Provided that no [food] additive shall be deemed to be safe if it is found to induce cancer when ingested by man or animal, or if it is found, after tests which are appropriate for the evaluation of the safety of food additives, to induce cancer in man or animal,…

From the earlier table, you can see that the Delaney Clause only applies to food additives, color additives, and new animal drug residues. That represents only three of the nine categories of food substances in the table. In other words, although the law prohibits any cancer-causing food additives from being used in food, it does not prevent cancer-causing substances that fit into one of the other six categories from being in food. We know, for example, that there are some naturally occurring cancer-causing substances in food. There are also some cancer-causing "added poisonous or deleterious substances" in food. The FDA can set a limit on how much of these substances should be allowed in food, but the law does not require that the FDA ban these substances from foods. However, in the case of a cancer-causing food additive, there is no allowable level. According to the Delaney Clause, if it is a food additive (or color additive or new animal drug) and if it is found to cause cancer, it is banned from use in foods.

Now, if we determine that the food additive does not cause cancer, then the second safety standard condition must still be satisfied. This states that in order for the food additive to be considered "safe," there must be "reasonable certainty of no harm" resulting from use of the substance. The legal basis for this safety standard is somewhat gray. When Congress passed the Food Additive Amendment in 1958, it did not fully appreciate the complexities associated with determining "safety." The amendment simply stated the following in this regard:

21 USC §348(c)(5)

In determining, for the purposes of this section, whether a proposed use of a food additive is safe, the Secretary shall consider among other relevant factors—

(A) the probable consumption of the additive and of any substance formed in or on food because of the use of the additive;

(B) the cumulative effect of such additive in the diet of man or animals, taking into account any chemically or pharmacologically related substance or substances in such diet; and

(C) safety factors, which in the opinion of experts qualified by scientific training and experience to evaluate the safety of food additives are generally recognized as appropriate for the use of animal experimentation data.

Thus, Congress outlined the factors that the FDA should consider when determining whether a food additive is safe, but it did not specify how those factors should be evaluated and weighed to make a determination of safety. That was left to the FDA. The agency addressed this in its regulations pertaining to food additive safety. Title 21 of the Code of Federal Regulations (CFR) includes the following:

21 CFR §170.3(i)

Safe or safety means that there is a reasonable certainty in the minds of competent scientists that the substance is not harmful under the intended conditions of use. It is impossible in the present state of scientific knowledge to establish with complete certainty the absolute harmlessness of the use of any substance. Safety may be determined by scientific procedures or by general recognition of safety.

This safety standard for food additives in the CFR is commonly abbreviated simply as "reasonable certainty of no harm." It is important to understand that this is a relatively strict standard. That is, even though it is impossible to establish "absolute harmlessness" for any substance, this standard tries to come as close as possible to this expectation by requiring "reasonable certainty" that the substance is "not harmful." It should not be surprising that the law and the regulations require such a high standard of safety for food additives, as they are not required in the manufacturing of a food. Rather, they are *optional* ingredients that a manufacturer may want to add to a food for a particular purpose. They should be held to a high safety standard.

One final point with regard to the safety of food additives: The burden of proof for establishing the safety of a new food additive lies with the manufacturer. That is, the manufacturer must provide the FDA with all of the scientific data in support of the food additive's safety. Until the FDA and its scientific review panels review these data and approve the use of the new food additive, it is not allowed for use in foods.

Now let us consider the safety standard for another category of food substances from the earlier table: *added poisonous or deleterious substances*. The term "added" does not necessarily mean that the manufacturer intentionally added a poisonous substance to the food. That would certainly be a serious crime. Rather, it typically describes substances that may be impossible to eliminate from foods entirely, but for which the manufacturer had some control over how much of the poisonous substance ended up in the food. Consider the example of the added poisonous or deleterious substance that I included in the table: aflatoxin. Aflatoxin is a very dangerous toxin that is produced by certain molds that may grow on some foods, such as peanuts. In addition to being a known cancer-causing substance, it can cause other serious health problems when consumed at levels that are considered "unsafe." Since aflatoxin is produced from a natural mold that can grow on foods, it may be tempting to consider it a "naturally occurring poisonous or deleterious substance." However, the farmer, processor, or food manufacturer has quite a bit of control over the conditions that allow the mold to grow and produce toxin. Individuals who handle these foods are expected to ensure that the food is grown, harvested, stored, and processed under conditions that limit or prevent mold growth. If the mold grows and produces toxin, the farmer, processor, manufacturer, etc. would be somewhat responsible. Therefore, the law considers it to be an *added poisonous or deleterious substance*.

It is interesting to note that even though aflatoxin is a known carcinogen (cancer-causing) substance, it is not banned from presence in food. That is because it is not a food additive, color additive, or animal drug residue (by legal definition), and therefore, the Delaney Clause does not apply. Furthermore, it would be impossible to eliminate all detectable levels of aflatoxin from food without possibly having to entirely eliminate the food itself. So the FD&C Act specifies a safety standard for added poisonous or deleterious substances as follows:

21 USC § 342(a)(1)
A food shall be deemed to be adulterated—
If it bears or contains any poisonous or deleterious substance which may render it injurious to health;

The key part of this phrase of the law is, *"may* render it injurious to health." Like the food additive safety standard (*reasonable certainty of no harm*), this is also a very strict safety standard. To declare a food adulterated (and therefore in violation of the FD&C Act), the FDA would simply have to conclude that the levels of the substance in the food *may render* the food injurious to health. In other words, the mere possibility of injury or harm would be enough to declare the food adulterated. Again, this should not be surprising as the manufacturer has some control over how much of the *added poisonous or deleterious substance* ends up in the food. Therefore, it should be held to a very high expectation of safety. Nevertheless, the FDA also recognizes that it may be impossible to completely eliminate any presence of the poisonous or deleterious substance in the food. So the FDA will set an acceptable level of the substance in the food, below which the food is considered "safe." This level would theoretically represent a level that would present as close to no possibility of harm as could reasonably be expected (given the fact that there is no such thing as zero risk). In the case of our aflatoxin example, the FDA has set an *action level* (upper limit) for aflatoxin in peanuts or peanut products of 20 parts per billion (ppb).

Finally, let's consider the third category of food substances: *naturally occurring poisonous or deleterious substances*. These are considered "not added" because they are naturally present in foods and the manufacturer has essentially no control over their presence. The section of the FD&C Act that deals with added poisonous or deleterious substances (quoted above) goes on to state the following italicized information related to "nonadded" poisonous or deleterious substances:

21 USC § 342(a)(1)
A food shall be deemed to be adulterated—
If it bears or contains any poisonous or deleterious substance which may render it injurious to health; *but in case the substance is not an added substance, such food shall not be considered adulterated under this clause if the quantity of such substance in such food does not ordinarily render it injurious to health.*

Thus, for *not added* poisonous or deleterious substances (such as naturally occurring substances), the safety standard is *ordinarily render it injurious to health*. In other words, the food would be considered adulterated if the level of the substance *is likely* to cause harm. This is a much weaker standard than the *may render injurious* standard that applies to added substances. Clearly, Congress recognized that since these substances are largely unavoidable to some extent in foods, it would not be feasible to apply a safety standard as strict as the food additive standard (*reasonable certainty of no harm*) or the added substances standard (*may render injurious*). Nevertheless, it is still a safety standard that should protect the public from foods that contain levels of the naturally occurring substance that present a likely risk of causing harm.

With the passage of the DSHEA in 1994, a tenth distinct category of food substances was created, namely *dietary supplements*. One of the provisions of this amendment to the FD&C Act was to create a new safety standard for dietary supplements.

---

21 USC §342(f)

A food shall be deemed to be adulterated—

(1) If it is a dietary supplement or contains a dietary ingredient that—

(A) presents a *significant or unreasonable risk of illness or injury* under—
    (i) conditions of use recommended or suggested in labeling, or
    (ii) if no conditions of use are suggested or recommended in the labeling, under ordinary conditions of use;
(B) is a new dietary ingredient for which there is inadequate information to provide reasonable assurance that such ingredient does not present a *significant or unreasonable risk of illness or injury;*

I have italicized the important phrasing in the above quote from the law… a *significant or unreasonable risk of illness or injury*. This is the new safety standard for dietary supplements as specified by the DSHEA. This is a very weak and ambiguous safety standard. How does one determine whether a risk is "unreasonable?" The FDA has been struggling with this since the DSHEA was enacted. To date, it has only found one dietary supplement ingredient to be "unsafe" within the meaning of this standard. In February of 2004, the FDA finalized a regulation pertaining to the weight loss dietary supplement ingredient, ephedra.

21 CFR § 119.1 Dietary supplements containing ephedrine alkaloids.

Dietary supplements containing ephedrine alkaloids present an unreasonable risk of illness or injury under conditions of use recommended or suggested in the labeling, or if no conditions of use are recommended or suggested in the labeling, under ordinary conditions of use. Therefore, dietary supplements containing ephedrine alkaloids are adulterated under section 402(f)(1)(A) of the Federal FD&C Act.

Two other important points to note about the safety standard that applies to dietary supplements. First, the burden of proof for determining whether a dietary supplement (or dietary supplement ingredient) presents a *significant or unreasonable risk of illness or injury* lies with the FDA, not the manufacturer. This effectively means that dietary supplements are presumed to be safe until the FDA can proof otherwise after the supplement is already on the market.

Second, the safety standard, as stated in 21 USC §342(f) requires that the FDA limit its determination as to whether or not the dietary supplement presents a *significant or unreasonable risk of illness or injury* to only "under the conditions of use recommended or suggested in labeling." Thus, if the label indicates that the supplement should only be taken at a dose of one pill per day, the FDA would only have the authority to evaluate and consider the risk at this dose, even though "ordinary use" of the supplement may be at a much higher dose. However, if the label does not recommend how the supplement should be used, then the FDA is free to consider "ordinary use" when evaluating risk.

All of this may seem to indicate that the FDA's authority to protect the public from potentially dangerous dietary supplements is severely restricted by the DSHEA. This is certainly true when you compare the FDA's relatively weak authority over dietary supplement safety to its much stronger authority over the safety of other food substances. However, supporters of the DSHEA are quick to point out that the law does give the FDA the authority to take quick action against dietary supplements or ingredients that it feels present an "imminent hazard to public health."

21 USC §342(f)

A food shall be deemed to be adulterated—

(1) If it is a dietary supplement or contains a dietary ingredient that—

(C) the Secretary declares to pose an imminent hazard to public health or safety, except that the authority to make such declaration shall not be delegated and the Secretary shall promptly after such a declaration initiate a proceeding in accordance with sections 554 and 556 of title 5 to affirm or withdraw the declaration;

While it is true that the FDA does have this authority under DSHEA, the "imminent hazard to public health and safety" is a high standard to meet. This provision of the law does little to protect public health and safety from possible dietary supplements or supplement ingredients that pose less severe risks, but nevertheless real risks that many health professionals would consider unnecessary and avoidable. Nor does it protect the public from the "indirect risks" of dietary supplement use, the effects of which are much more difficult to measure. As Dr. Kessler noted in his Senate hearing testimony on S. 784, these indirect risks refer to the health risks associated with relying on supplements to treat or prevent serious illnesses that would be more effectively handled by well-tested and safe mainstream medical treatments.

chapter twelve

DSHEA

Structure–function claims versus health claims

Based on much of what we have covered in earlier chapters of this book, it should now be quite clear that one of the most contentious and controversial issues driving passage of the Dietary Supplement Health and Education Act (DSHEA) was the issue of health claims and how the Food and Drug Administration (FDA) chose to enforce the health claim provisions of the 1990 Nutrition Labeling and Education Act (NLEA) with regard to dietary supplements. In the early years following the passage of the NLEA, the FDA made it very clear that it intended to hold dietary supplements that were seeking to make a disease-related health claim to the same "significant scientific agreement" standard that it required of conventional foods. This was in spite of the fact that when Congress passed the NLEA, it had granted the FDA the option of establishing a different, perhaps less stringent standard for dietary supplements. By sticking with the *significant scientific agreement* standard, the FDA was effectively preventing dietary supplements from making explicit health claims on their labeling. To date, the only two dietary supplement health claims that are approved by the FDA are the "calcium and osteoporosis" health claim and the "folic acid and neural tube defect" health claim. As far as the FDA is concerned, no other dietary supplement and disease relationship has met the "significant scientific agreement" requirement.

While some of the early drafts of the bill that would ultimately become the DSHEA did grant dietary supplement manufacturers more freedom to make explicit health claims, the final version of the law left this issue largely the same as it was in the NLEA. That is, the FDA would continue to apply the same "significant scientific agreement" standard to both conventional foods and dietary supplements. Any product labeling reference to a disease claim that was not included among the few FDA approved health claims would render the product an "unapproved drug" rather than a food or dietary supplement. However, the DSHEA did provide for some very important new labeling options for dietary supplements. Unlike the explicit health claims discussed earlier, these claims would not require preapproval by the FDA.

21 U.S. code (USC) § 343(r)(6) For purposes of paragraph (r)(1)(B), a statement for a dietary supplement may be made if—

(A) the statement claims a benefit related to a classical nutrient deficiency disease and discloses the prevalence of such disease in the United States, describes the role of a nutrient or dietary ingredient intended to affect the structure or function in humans, characterizes the documented mechanism by which a nutrient or dietary ingredient acts to maintain such structure or function, or describes general well-being from consumption of a nutrient or dietary ingredient,

(B) the manufacturer of the dietary supplement has substantiation that such statement is truthful and not misleading, and

(C) the statement contains, prominently displayed and in boldface type, the following: "This statement has not been evaluated by the FDA. This product is not intended to diagnose, treat, cure, or prevent any disease."

A statement under this subparagraph may not claim to diagnose, mitigate, treat, cure, or prevent a specific disease or class of diseases. If the manufacturer of a dietary supplement proposes to make a statement described in the first sentence of this subparagraph in the labeling of the dietary supplement, the manufacturer shall notify the Secretary no later than 30 days after the first marketing of the dietary supplement with such statement that such a statement is being made.

Thus, Congress allowed for three possible "claims" on the labeling of dietary supplements. One, a claim that conveys the role of a nutrient in preventing a deficiency disease. Two, a claim that describes the role of a nutrient or dietary ingredient on the structure or function in the body. And three, a claim that describes the role that a nutrient or dietary ingredient can play in the general well-being of an individual.

This section of the law specifies three things a dietary supplement manufacturer is required to do if it wants to include one of these three types of claims on its product labeling.

1. The manufacturer is required to have substantiation that the claim is truthful and not misleading. Note that the law does not require that the manufacturer provides this substantiation information to the FDA. It simply requires that the manufacturer has this information.

Remember, there is no FDA preapproval requirement for these types of claims. The FDA has prepared a "Guidance for Industry" document that provides detailed information on what the FDA would consider appropriate substantiation evidence.[1]

2. The manufacturer is required to notify the FDA that the product includes one of these three claims on its labeling no later than 30 days after the product is released on the market.

3. The manufacturer must include the following disclaimer on the labeling: "This statement has not been evaluated by the Food and Drug Administration. This product is not intended to diagnose, treat, cure, or prevent any disease."

Of the three permitted dietary supplement labeling claims listed in 21 USC §343(r)(6) earlier, the one that is certainly utilized and exploited to the greatest extent by the dietary supplement industry is the structure/function claim. It is this claim option that allows manufacturers to get as close as possible to linking their supplement products to specific health conditions, while avoiding making a "disease claim" that would trigger the drug definition of the law. Anyone who has ever perused the selves of a dietary supplement store or scanned the print ads for dietary supplements in health magazines can appreciate the very fine line that exists between a legitimate structure/function claim and an illegal disease claim. This has long been one of the FDA's major concerns regarding the marketing of dietary supplements over the years and certainly one of its major concerns with the DSHEA. Unfortunately, the DSHEA did not provide much detail for what would constitute a legal structure/function claim. The 21 USC §343(r)(6) simply states that *"a statement for a dietary supplement may be made if [it]… describes the role of a nutrient or dietary ingredient intended to affect the structure or function in humans, [or] characterizes the documented mechanism by which a nutrient or dietary ingredient acts to maintain such structure or function…"* It was thus up to the FDA to draft regulations that would provide the necessary details and guidance for constructing structure/function claims that meet the requirements of the law. These regulations can be found in 21 Code of Federal Regulations (CFR) §101.93. The regulations start by providing a slightly more expanded definition of a structure/function claim:

---

[1] *Guidance for Industry: Substantiation for Dietary Supplement Claims Made Under Section 403(r) (6) of the Federal Food, Drug, and Cosmetic Act.* U.S. Food and Drug Administration. www. fda.gov/food/guidanceregulation/guidancedocumentsregulatoryinformation/dietary-supplements/ucm073200.htm.

21 CFR §101.93(f)
Permitted structure/function statements
Dietary supplement labels or labeling may, subject to the requirements in paragraphs (a) through (e) of this section, bear statements that describe the role of a nutrient or dietary ingredient intended to affect the structure or function in humans or that characterize the documented mechanism by which a nutrient or dietary ingredient acts to maintain such structure or function, provided that such statements are not disease claims under paragraph (g) of this section. If the label or labeling of a product marketed as a dietary supplement bears a disease claim as defined in paragraph (g) of this section, the product will be subject to regulation as a drug unless the claim is an authorized health claim for which the product qualifies.

This section is important because it makes it clear that structure/function claims are permitted as long as they do not make a "disease claim." The regulations then go on to define the term "disease":

21 CFR §101.93(g)(1)

For purpose of 21 USC 343(r)(6), a "disease" is damage to an organ, part, structure, or system of the body, such that it does not function properly (e.g., cardiovascular disease) or a state of health leading to such dysfunctioning (e.g., hypertension); except that diseases resulting from essential nutrient deficiencies (e.g., scurvy, pellagra) are not included in this definition.

In drafting these regulations, the FDA clearly felt that to help manufacturers distinguish between a structure/function claim and a disease claim, it would be easier to focus on the criteria that would suggest a disease claim. If the claim does not meet one of the criteria for a disease claim, then it would likely be an allowable structure/function claim. Thus, the remainder of 21 CFR 101.93 outlines the criteria for meeting the definition of a disease claim.

The ten criteria for a disease claim included in §101.93 earlier cover a very wide range of considerations. Recall that the legal definition of a drug states, *"The term drug means... articles intended for use in the diagnosis, cure, mitigation, treatment, or prevention of disease in man or other animals."*

21 CFR §101.93(g)(2)

FDA will find that a statement about a product claims to diagnose, mitigate, treat, cure, or prevent disease (other than a classical nutrient deficiency disease) under 21 USC 343(r)(6) if it meets one or more of the criteria listed below. These criteria are not intended to classify as disease claims statements that refer to the ability of a product to maintain healthy structure or function, unless the statement implies disease prevention or treatment. In determining whether a statement is a disease claim under these criteria, FDA will consider the context in which the claim is presented. A statement claims to diagnose, mitigate, treat, cure, or prevent disease if it claims, explicitly or implicitly, that the product:

(i) Has an effect on a specific disease or class of diseases;

(ii) Has an effect on the characteristic signs or symptoms of a specific disease or class of diseases, using scientific or lay terminology;

(iii) Has an effect on an abnormal condition associated with a natural state or process, if the abnormal condition is uncommon or can cause significant or permanent harm;

(iv) Has an effect on a disease or diseases through one or more of the following factors:

(A) The name of the product;

(B) A statement about the formulation of the product, including a claim that the product contains an ingredient (other than an ingredient that is an article included in the definition of "dietary supplement" under 21 USC 321(ff)(3)) that has been regulated by FDA as a drug and is well known to consumers for its use or claimed use in preventing or treating a disease;

(C) Citation of a publication or reference, if the citation refers to a disease use, and if, in the context of the labeling as a whole, the citation implies treatment or prevention of a disease, e.g., through placement on the immediate product label or packaging, inappropriate prominence, or lack of relationship to the product's express claims;

(D) Use of the term "disease" or "diseased," except in general statements about disease prevention that do not refer explicitly or implicitly to a specific disease or class of diseases or to a specific product or ingredient; or

(E) Use of pictures, vignettes, symbols, or other means;

(v) Belongs to a class of products that is intended to diagnose, mitigate, treat, cure, or prevent a disease;

(vi) Is a substitute for a product that is a therapy for a disease;

(vii) Augments a particular therapy or drug action that is intended to diagnose, mitigate, treat, cure, or prevent a disease or class of diseases;

(viii) Has a role in the body's response to a disease or to a vector of disease;

(ix) Treats, prevents, or mitigates adverse events associated with a therapy for a disease, if the adverse events constitute diseases; or

(x) Otherwise suggests an effect on a disease or diseases.

An important key word in this phrase is *intended*. To be considered a disease claim, it is not necessary for the label to make an explicit link between a supplement and a particular disease (such as, *Supplement X may reduce your risk of cancer*). It is only necessary that the claim convey or imply that the *intended* purpose of the product is to diagnose, cure, mitigate, treat, or prevent a disease. Thus, the FDA's ten criteria for a disease claim extend to considerations such as the impression that is made by the name of the supplement, or pictures or symbols associated with the supplement, or any association between the supplement and the symptoms of a disease, etc.

Some proponents and advocates in the use of dietary supplements, and those who are involved in their marketing and sales, may have been disappointed that the DSHEA did not grant them more lenient "heath and disease claim" options in their labeling. However, the DSHEA's structure–function labeling options have proven to be a rather powerful labeling and advertising tool. To be in compliance with these labeling options, it is important that dietary supplement manufacturers clearly understand the limits of these allowable claims. To assist manufacturers in this process, the FDA has prepared a Compliance Guide for Industry which provides much more detail on the ten "disease claim" criteria that must be avoided in an allowable structure–function claim.[2]

---

[2] *Guidance for Industry: Structure/Function Claims, Small Entity Compliance Guide.* U.S. Food and Drug Administration. www.fda.gov/Food/GuidanceRegulation/GuidanceDocumentsRegulatoryInformation/DietarySupplements/ucm103340.htm.

chapter thirteen

DSHEA

Other important provisions

The Dietary Supplement Health and Education Act (DSHEA) had some other very important provisions, including significant changes to the labeling and manufacturing of these products, as well as provisions to encourage research and education related to the value of dietary supplements. We'll first examine the labeling provisions of the DSHEA. Recall from Chapter six of the Food, Drug, and Cosmetic (FD&C) Act, the definition of the terms "label" and "labeling."

> **21 The U.S. code (USC) § 321: *Definitions*.**
>
> **(k) The term "label" means a display of written, printed, or graphic matter upon the immediate container of any article; ...**
>
> **(m) The term "labeling" means all labels and other written, printed, or graphic matter (1) upon any article or any of its containers or wrappers or (2) accompanying such article.**

"Accompanying such article" in the *labeling* definition was typically interpreted as meaning that it was displayed in close proximity to the product (article) for the intended purpose of having consumers link the two. Based on these definitions, a pamphlet or article would be considered part of the "labeling" of a product if the pamphlet was "accompanying" the product. Before passage of the DSHEA, if the pamphlet or article met any of the criteria for a disease claim (similar to those specified in 21 Code of Federal Regulations (CFR) §101.93(g)(2)), then the *labeling* of the product would be considered to include the disease claim and the Food and Drug Administration (FDA) could declare that the product met the legal definition of a drug.

For example, let us suppose that a research article is published in the *New England Journal of Medicine* that reported the results of a study on the effects of omega-3 fatty acids on inflammation in the blood vessels of rats. Inflammation in blood vessels is known to be a factor in the process of atherosclerosis. Thus, the focus of the article could be construed as meeting the criteria for a "disease claim" even though the investigators/authors of the study had no intention to suggest that it be used for this purpose. If the owner of a health food store were to place copies of this article next to bottles of fish oil supplements for sale, the journal article would now meet the legal definition as part of the "labeling" of the product. Since the labeling now includes a disease claim (the article), the FDA could consider the fish oil supplement itself as meeting the legal definition of a drug.

Before the passage of the DSHEA in 1994, the FDA used this enforcement strategy to prevent individuals who sold dietary supplements from using this type of so-called "third-party literature" to help sell their products. Dietary supplement manufacturers and marketers felt that this was a violation of their protected free speech rights and succeeded in getting Congress to address this issue. The DSHEA amended the FD&C Act in this regard as follows:

21 USC §343–2. Dietary supplement labeling exemptions

(a) In general

A publication, including an article, a chapter in a book, or an official abstract of a peer-reviewed scientific publication that appears in an article and was prepared by the author or the editors of the publication, which is reprinted in its entirety, shall not be defined as labeling when used in connection with the sale of a dietary supplement to consumers when it—

(1) is not false or misleading;

(2) does not promote a particular manufacturer or brand of a dietary supplement;

(3) is displayed or presented, or is displayed or presented with other such items on the same subject matter, so as to present a balanced view of the available scientific information on a dietary supplement;

(4) if displayed in an establishment, is physically separate from the dietary supplements; and

(5) does not have appended to it any information by sticker or any other method.

Note that this new provision of the law does not give the dietary supplement industry total free rein to include any third-party literature.

However, if the third-party literature is not false or misleading, does not promote a particular product, is included as part of a balanced view, is physically separate from the product, and does not have anything appended to it (presumably that might promote the product), then it would be exempt from the definition of "labeling."

The "Supplement Facts" label

The DSHEA included some new requirements for the nutrition labeling of dietary supplements. Recall that the Nutrition Labeling and Education Act of 1990 mandated nutrition labeling for most food products on the market, including dietary supplements. However, many of the specific requirements for the nutrition labeling of conventional foods did not fit well with the unique nature and formulations of typical dietary supplement products. The DSHEA labeling requirements addressed many of these unique considerations.

21 U.S. code (USC) §343(q)(5)(F)
A dietary supplement product (including a food to which section 350 of this title applies) shall comply with the requirements of subparagraphs (1) and (2) in a manner which is appropriate for the product and which is specified in regulations of the Secretary which shall provide that—

(i) nutrition information shall first list those dietary ingredients that are present in the product in a significant amount and for which a recommendation for daily consumption has been established by the Secretary, except that a dietary ingredient shall not be required to be listed if it is not present in a significant amount, and shall list any other dietary ingredient present and identified as having no such recommendation;

(ii) the listing of dietary ingredients shall include the quantity of each such ingredient (or of a proprietary blend of such ingredients) per serving;

(iii) the listing of dietary ingredients may include the source of a dietary ingredient; and

(iv) the nutrition information shall immediately precede the ingredient information required under subclause (i), except that no ingredient identified pursuant to subclause (i) shall be required to be identified a second time.

As is typical with labeling issues, the Congress left most of the details to the enforcement agency ("the Secretary" of Health and Human Services who would delegate it to the FDA). But it did lay out the basic intent of the labeling for dietary supplements. That is, the nutrition information should be "in a manner which is appropriate for the product." This allowed the FDA to tailor the format and requirements to the unique characteristics of dietary supplements. The law also specifies that the nutrition information should include only those dietary ingredients that are present in a significant amount, the quantity of each such ingredient, the option to include the source of the dietary ingredient, and the requirement that other "non-dietary" ingredients be listed immediately following the nutrition information. The FDA's final detailed regulations enforcing this section of the law can be found in the Code of Federal Regulations (CFR) at 21 CFR §101.36. These regulations specific to the nutrition labeling of dietary supplements cover 14 pages in the CFR, and it is not necessary to discuss them all in detail here. However, the most immediately noticeable difference between the nutrition labeling of conventional foods and that for dietary supplements is the heading on the label. We are all familiar with the "Nutrition Facts" panel on conventional foods. Dietary supplements use a "Supplement Facts" panel. The CFR includes several sample formats for this panel. I have included one below.

Supplement Facts

Serving Size 1 Capsule

Amount Per Capsule		% Daily Value
Calories 20		
Calories from Fat	20	
Total Fat 2 g		3% *
Saturated Fat 0.5 g		3% *
Polyunsaturated Fat 1 g		†
Monounsaturated Fat 0.5 g		†
Vitamin A 4250 IU		85%
Vitamin D 425 IU		106%
Omega-3 Fatty Acids 0.5 g		†

* Percent Daily Values are based on a 2,000 calorie diet.
† Daily Value not established.

Ingredients: Cod liver oil, gelatin, water, and glycerin.

The FDA has prepared a "Dietary Supplement Labeling Guide" to provide further guidance in preparing "Supplement Facts" labels that are in compliance with the law and associated regulations.[1] As part of this guide, the FDA identifies the essential difference between a "Supplements Facts" panel and a "Nutrition Facts" panel.

The major differences between "Supplement Facts" panel and "Nutrition Facts" panel are as follows:

a. You must list dietary ingredients without Reference Daily Intakes (RDIs) or Daily Reference Values (DRVs) in the "Supplement Facts" panel for dietary supplements. You are not permitted to list these ingredients in the "Nutrition Facts" panel for foods.

b. You may list the source of a dietary ingredient in the "Supplement Facts" panel for dietary supplements. You cannot list the source of a dietary ingredient in the "Nutrition Facts" panel for foods.

c. You are not required to list the source of a dietary ingredient in the ingredient statement for dietary supplements if it is listed in the "Supplement Facts" panel.

d. You must include the part of the plant from which a dietary ingredient is derived in the "Supplement Facts" panel for dietary supplements. You are not permitted to list the part of a plant in the "Nutrition Facts" panel for foods.

e. You are not permitted to list "zero" amounts of nutrients in the "Supplement Facts" panel for dietary supplements. You are required to list "zero" amounts of nutrients in the "Nutrition Facts" panel for food.

From: Dietary Supplement Labeling Guide: Chapter IV. Nutrition Labeling, U.S. Food and Drug Administration, 2005. www.fda.gov/Food/GuidanceRegulation/ GuidanceDocumentsRegulatoryInformation/DietarySupplements/ ucm070597.htm.

---

[1] *Dietary Supplement Labeling Guide.* U.S. Food and Drug Administration. www.fda. gov/Food/GuidanceRegulation/GuidanceDocumentsRegulatoryInformation/ DietarySupplements/ucm2006823.htm.

Current good manufacturing practices for dietary supplements

The FDA has long used its regulatory authority to establish current good manufacturing practices (CGMPs) that food manufacturers must follow with the principle goal of enforcing the food sanitation and food safety provisions of the FD&C Act. CGMPs typically cover all aspects of the harvesting, processing, transportation, storage, distribution and more of foods, as well as personnel hygiene, record keeping, sampling, and testing. The goal of food CGMPs is for food manufacturers to establish operating procedures that ensure that all foods are safe and properly labeled. Most food manufacturers can follow the general (or sometimes referred to as the "umbrella") good manufacturing practices (GMPs) that specify the operating procedures that the manufacturers must follow, regardless of what particular food they are producing. However, some foods present unique safety and/or sanitation issues and risks that require specific GMPs for that particular food. For example, manufacturers of infant formula must follow CGMPs specific for that product. With the passage of the DSHEA in 1994, Congress provided for the FDA to establish CGMPs specifically developed for the dietary supplement industry.

---

21 USC §342(g): Dietary supplement: manufacturing practices

A food shall be deemed to be adulterated—

(1) If it is a dietary supplement and it has been prepared, packed, or held under conditions that do not meet current good manufacturing practice regulations, including regulations requiring, when necessary, expiration date labeling, issued by the Secretary under subparagraph (2).

(2) The Secretary may by regulation prescribe good manufacturing practices for dietary supplements. Such regulations shall be modeled after current good manufacturing practice regulations for food and may not impose standards for which there is no current and generally available analytical methodology. No standard of current good manufacturing practice may be imposed, unless such standard is included in a regulation promulgated after notice and opportunity for comment in accordance with Chapter 5 of title 5.

The phrasing of this provision of the law is intentionally broad, deferring to the FDA (as delegated by the Secretary of HHS) to provide all of the necessary CGMP details to enforce the law. Although the DSHEA amendment to the FD&C Act was passed in 1994, it was not until 2007 that the FDA published its final Dietary Supplement CGMP regulations in the Federal Register (72 FR 34752). These were then codified in 21 CFR Part 111.

Like the "umbrella" CGMPs for most foods, the dietary supplement CGMPs include all of the requirements necessary to ensure that the products are not contaminated with microorganisms, pesticides, or other potentially dangerous substances. But the FDA was also acutely aware that there were other potential problems with some dietary supplements. For example, some dietary supplement manufacturers were selling products that did not contain the dietary ingredient at the level indicated on the label. In some cases, products were being sold, which did not contain any of the dietary ingredient on the label. In an FDA "backgrounder" document to its final dietary supplement CGMPSs, the agency explained how its new regulations would address this.

The new *dietary supplement CGMPs:*

Requires certain activities in manufacturing, packaging, labeling, and holding of dietary supplements to ensure that a dietary supplement contains what it is labeled to contain and is not contaminated with harmful or undesirable substances such as pesticides, heavy metals, or other impurities.

Requires certain activities that will ensure the identity, purity, quality, strength, and composition of dietary supplements, which is a significant step in assuring consumers they are purchasing the type and amount of ingredients declared.

Backgrounder: Final Rule for Current Good Manufacturing practices (CGMPs) for Dietary Supplements. U.S. Food and Drug Administration. June, 2007.

The new CGMP regulations do not specify exactly which scientific test needs to be used to *ensure the identity and purity* of a particular dietary supplement ingredient. The FDA recognized, for example, that a method that would be used to test for the identity and purity of a vitamin in a supplement would certainly be different from a method for a herbal ingredient in a supplement. Therefore, the regulations simply require that a "scientifically valid method" be used that is "accurate, precise, and specific for its intended purpose."

There is still some controversy regarding what is the most appropriate "scientifically valid method" for different dietary supplement ingredients. This controversy was recently and notably played out in the popular press, when in February of 2015, the Attorney General (AG) for the State of New York issued cease and desist orders (to halt sales) against four retailers (GNC, Target, Walmart, and Walgreens). The order accused the retailers of selling store brand herbal supplements that did not contain the herbal ingredients listed on the label and/or that in some cases contained other plant ingredients not listed on the label.

According to the press release on the AG's website,[2] the cease and desist letters resulted from a study commissioned by the AG that found that, of 78 herbal supplement products tested from these retailers, DNA barcode testing found that only 21% contained the herbal ingredient indicated on the label (saw palmetto, Echinacea, ginseng, garlic, ginkgo biloba, St John's Wort). For products sold at Walmart, only 4% were found to contain the DNA from the herbal ingredient listed on the label. In addition, the DNA barcode testing revealed that 35% of the products tested contained DNA from plant species not listed on the ingredient labels, including rice, beans, pine, citrus, asparagus, primrose, wheat, houseplant, wild carrot, and more.

The controversy stems from the DNA barcode testing used in the study. DNA barcoding analyzes a small piece of an organism DNA and compares it to the DNA barcodes from a database of animal and plant species to identify the species in question. It is a highly sensitive and specific test. But it does have some important limitations. The dietary supplement industry argued that the New York AG's results may have been due to the degradation or destruction of the DNA in the herbal supplement products during processing, resulting in the lack of detectable DNA in the final tested sample. Nevertheless, the United States Pharmacopeia (USP), a science-based nonprofit organization that establishes standards for analysis and quality for foods, drugs, and dietary supplements, suggested that DNA barcode testing should be added to their battery of identification tests for plant materials in dietary supplements, with some qualifications:

---

[2] Press Release: *A.G. Schneiderman Asks Major Retailers To Halt Sales Of Certain Herbal Supplements As DNA Tests Fail To Detect Plant Materials Listed On Majority Of Products Tested.* https://ag.ny.gov/press-release/ag-schneiderman-asks-major-retailers-halt-sales-certain-herbal-supplements-dna-tests.

> *With the appropriate validation, USP may incorporate DNA tests into specific product quality standards, but even then it is envisioned not as stand-alone procedure, but as a complement to existing chromatographic, spectroscopic, and botanical morphological (microscopic/macroscopic) analytical procedures.*
>
> *DNA testing poses some unique benefits. Its sensitivity and specificity helps in accurate identification of the botanical species as well as adulterants and contaminants – especially in material where the macro-botanical characteristics are no longer present, such as powdered or ground material. However, this same sensitivity can lead to false results and that calls for careful interpretation of the data.*
>
> *Also, identity is just one of many attributes that are used to determine the overall quality of a given material. Quality of plant materials is determined by identity, purity and the content of bioactive constituents. DNA based tests should not be used as the only way to determine botanical content.*
>
> *There are also materials for which DNA testing is not recommended. DNA-based methods are not suitable for materials, such as botanical extracts, that were subjected to processes that denature, degrade and destroy DNA.*
>
> *The United States Parmacopeial Convention Urges Scientific Validation of DNA Test Methods for Regulating the Quality of Herbal Supplements.* USP Statement. April 16, 2015.
> www.usp.org/news/united-states-pharmacopeial-convention-urges-scientific-validation-dna-test-methods-regulating-quality-herbal-supplements.

Thus, the USP recognizes the value of DNA barcode identification testing, but only when used with other confirming tests, such as chromatographic, spectroscopic, and macro/microscopic tests. In the meantime, the New York State AG's Office has reached agreements with GNC, Nature's Way, and NBTY (formerly known as Nature's Bounty and the manufacturer of herbal supplements for the Walmart and Walgreen's label), to improve quality control procedures and to incorporate DNA barcode testing into its battery of analytical testing methods. However, the FDA's CGMPs for dietary supplements continues to only require a "scientifically valid method" for identity and quality testing. It is, nevertheless, incumbent on the manufacturers to meet the requirements of the CGMPs, regardless of what methods they use to ensure this.

The Office of Dietary Supplements

The last major provision of the DSHEA that I will briefly cover is the establishment of an Office of Dietary Supplements (ODS) within the National Institutes of Health (NIH). Since this Office is part of the NIH, and not part of the FDA, the provision of the law that established this Office is found in Title 42 of the U.S. code (USC) (covering laws related to various aspects of Public Health and Welfare), rather than Title 21 (where most FDA laws are codified).

---

42 USC §287c-11 Dietary supplements

(a) Establishment
The Secretary shall establish an Office of Dietary Supplements within the National Institutes of Health.
(b) Purpose
The purposes of the Office are—
>   (1) to explore more fully the potential role of dietary supplements as a significant part of the efforts of the United States to improve healthcare and
>   (2) to promote scientific study of the benefits of dietary supplements in maintaining health and preventing chronic disease and other health-related conditions.

(c) Duties
The Director of the Office of Dietary Supplements shall—
>   (1) conduct and coordinate scientific research within the National Institutes of Health relating to dietary supplements and the extent to which the use of dietary supplements can limit or reduce the risk of diseases such as heart disease, cancer, birth defects, osteoporosis, cataracts, or prostatism;
>   (2) collect and compile the results of scientific research relating to dietary supplements, including scientific data from foreign sources or the Office of Alternative Medicine;
>   (3) serve as the principal advisor to the Secretary and to the Assistant Secretary for Health and provide advice to the Director of the National Institutes of Health, the Director of the Centers for Disease Control and Prevention, and the Commissioner of Food and Drugs on issues relating to dietary supplements including—
>>   (A) dietary intake regulations;
>>   (B) the safety of dietary supplements;
>>   (C) claims characterizing the relationship between—

(i) dietary supplements; and
(ii)(I) prevention of disease or other health-related conditions; and
(II) maintenance of health; and
(D) scientific issues arising in connection with the labeling and composition of dietary supplements;
(4) compile a database of scientific research on dietary supplements and individual nutrients; and
(5) coordinate funding relating to dietary supplements for the National Institutes of Health.

It is clear from the wording of this provision of the DSHEA that Congress believed that more federal research funding should be focused on the potential health benefits of dietary supplements. Note the strength of the language the Congress chose for describing the intended "purpose" of the new office. *"...to explore more fully the potential role of dietary supplements as a significant part of the efforts of the United States to improve health care."* And, *"...to promote scientific study of the benefits of dietary supplements in maintaining health and preventing chronic disease."* This language clearly conveys the sense that the Congress expected (or at least hoped) that this Office would generate scientific data that would support the health benefits of dietary supplements. Congress also clearly felt that the best way to accomplish this was by creating a new office within the NIH, for which this would be its primary focus and responsibility. It would have made little sense to establish this new office within the FDA, as the FDA is not a research organization. Its primary function is enforcement. The NIH's primary function is biomedical research.

The ODS is a relatively small part of the NIH. Its fiscal year 2016 budget was $25.3 million. While not a trivial amount of money, it pales in comparison to other Institutes and Centers within the NIH. For example, the FY 2016 budget for the National Cancer Institute (NCI) was nearly $5.1 billion, the National Institute of Allergy and Infectious Diseases was $4.6 billion, and the National Heart, Lung, and Blood Institute was nearly $3.1 billion. Even the National Institute for Complementary and Alternative Medicine FY 2016 budget was $128 million. But it is important to note that ODS accomplishes its research mission primarily by co-funding research projects with other Institutes and Centers within NIH. Thus, for example, a research project that seeks to explore the relationship between a particular vitamin and cancer may be funded primarily through the NCI with some amount of co-funding from ODS. An important role for the ODS, therefore, is to advocate for funding of these projects through other parts of the NIH. Of the ODS 2016 budget of $25.3 million, approximately 45% is spent on these co-funded research projects.

The ODS recently released its 5-year (2017–2021) strategic plan, identifying four major goals:

1. Expand the scientific knowledge base on dietary supplements by stimulating and supporting a full range of biomedical research and by developing and contributing to collaborative initiatives, workshops, meetings, and conferences.

2. Enhance the dietary supplement research workforce through training and career development.

3. Foster development and dissemination of research resources and tools to enhance the quality of dietary supplement research.

4. Translate dietary supplement research findings into useful information for consumers, health professionals, researchers, and policymakers.

NIH Office of Dietary Supplements. Strategic Plan 2017–2021.
Strengthening Knowledge & Understanding of Dietary Supplements.
December 2016. https://ods.od.nih.gov/pubs/strategicplan/
ODSStrategicPlan2017-2021.pdf.

The first three goals are primarily focused on its research mission. However, the fourth goal is particularly important. Consumers, health professionals, researchers, and policymakers need to be able to distinguish between scientifically sound and creditable information on dietary supplements, and the confusing and often exaggerated claims on product labeling and advertising. The ODS accomplishes this goal in a number of ways. One of the most useful is its online "Dietary Supplement Fact Sheets[3]." Here both consumers and health professionals can find the latest information on the safety and effectiveness of more than 130 dietary supplement substances. Browsing through these fact sheets is a very enlightening exercise. Despite more than two decades since the passage of the DSHEA and despite the Congress' intended purpose for ODS to "promote scientific study of the benefits of dietary supplements," the fact

---

[3] *Dietary Supplement Fact Sheets*. Office of Dietary Supplement. National Institutes of Health. https://ods.od.nih.gov/factsheets/list-all/.

sheets overwhelmingly show at best, very weak or inconclusive evidence of benefits for the vast majority of dietary supplements. In many cases, the scientific data explained in the fact sheets suggest potentially significant risks from some dietary supplements. Consider this excerpt from the vitamin E fact sheet.

What are some effects of vitamin E on health?

Heart disease

Some studies link higher intakes of vitamin E from supplements to lower chances of developing heart disease. But the best research finds no benefit. People in these studies are randomly assigned to take vitamin E or a placebo (dummy pill with no vitamin E or active ingredients) and they don't know which they are taking. Vitamin E supplements do not seem to prevent heart disease, reduce its severity, or affect the risk of death from this disease. Scientists do not know whether high intakes of vitamin E might protect the heart in younger, healthier people who do not have a high risk of heart disease.

Cancer

Most research indicates that vitamin E does not help prevent cancer and may be harmful in some cases. Large doses of vitamin E have not consistently reduced the risk of colon and breast cancer in studies, for example. A large study found that taking vitamin E supplements (400 IU/day) for several years increased the risk of developing prostate cancer in men. Two studies that followed middle-aged men and women for 7 or more years found that extra vitamin E (300–400 IU/day, on average) did not protect them from any form of cancer. However, one study found a link between the use of vitamin E supplements for 10 years or more and a lower risk of death from bladder cancer.

Vitamin E dietary supplements and other antioxidants might interact with chemotherapy and radiation therapy. People undergoing these treatments should talk with their doctor or oncologist before taking vitamin E or other antioxidant supplements, especially in high doses.

Eye disorders

Age-related macular degeneration (AMD), or the loss of central vision in older people, and cataracts are among the most common causes of vision loss in older people. The results of research on whether vitamin E can help prevent these conditions are inconsistent. Among people with AMD who were at high risk of developing advanced AMD, a supplement containing large doses of vitamin E combined with other antioxidants, zinc, and copper showed promise for slowing down the rate of vision loss.

Mental function

Several studies have investigated whether vitamin E supplements might help older adults remain mentally alert and active as well as prevent or slow the decline of mental function and Alzheimer's disease. So far, the research provides little evidence that taking vitamin E supplements can help healthy people or people with mild mental functioning problems to maintain brain health.

Can vitamin E be harmful?

Eating vitamin E in foods is not risky or harmful. In supplement form, however, high doses of vitamin E might increase the risk of bleeding (by reducing the blood's ability to form clots after a cut or injury) and of serious bleeding in the brain (known as hemorrhagic stroke). Because of this risk, the upper limit for adults is 1,500 IU/day for supplements made from the natural form of vitamin E and 1,100 IU/day for supplements made from synthetic vitamin E. The upper limits for children are lower than those for adults. Some research suggests that taking vitamin E supplements even below these upper limits might cause harm. In one study, for example, men who took 400 IU of vitamin E each day for several years had an increased risk of prostate cancer.

   I think it is safe to conclude from this fact sheet that any possible (but still inconclusive) benefits of vitamin E supplements are far outweighed by the potential risks. This is a very important message to convey to the many millions of devoted dietary supplement users in the United States. However, I suspect that very few of them have ever checked the ODS website for information on the supplements they are taking, or are even aware that this website exists. Thus, despite the efforts of the ODS to organize, interpret, and provide this information, it is unclear how effective and successful the effort has been. In fact, the Pew Charitable Trust, an independent, nonpartisan research and policy organization, sent a letter to ODS in response to the draft of the ODS's most recent 5-year strategic plan. In the letter, the Pew organization strongly recommended that the

ODS expands its "Fourth Goal" to include *"evaluate the effectiveness of those resources in influencing consumer decisions and health care professionals' recommendations*[4]*."* Unfortunately, this recommendation was not included in the final ODS strategic plan document.

There have been various additional amendments to the FD&C Act since 1994 that have had some impact on the regulation of dietary supplements. In addition, there have been some important judicial decisions that have also impacted dietary supplement regulation. These "post-DSHEA" activities and events will be covered in the next chapter.

---

[4] The Pew Charitable Trust. *Letter to Paul M. Coates, Ph.D. Director, Office of Dietary Supplements, National Institutes of Health.* www.pewtrusts.org/~/media/assets/2016/10/pew_supports_office_of_dietary_supplements_strategic_plan_but_suggests_evaluation.pdf.

chapter fourteen

The health claims
debate continues

With the passage of the Nutrition Labeling and Education Act (NLEA) in 1990 and the Dietary Supplement Health and Education Act (DSHEA) in 1994, it would have appeared that the rules for what types of claims could be made on the labeling of dietary supplements was settled. The NLEA resolved the *explicit health claims* issue that began with the Kellogg's All-Bran ad campaign. Only those specific health/disease claims that were preapproved by the Food and Drug Administration (FDA) would be allowed on food labels, including dietary supplements. Under the FDA's enforcement of the NLEA, dietary supplements were excluded from making explicit health/disease claims on their labeling, with a few exceptions. Any food manufacturer, including dietary supplement manufacturers, could formally petition the FDA to have a new explicit health/disease claim approved for use on labeling. But the FDA would require that there be "significant scientific agreement" among qualified experts before it would approve any new claim. This was a very high standard to meet. In addition, for many years, the FDA did not even specify the criteria it would consider when determining whether the "significant scientific agreement" standard was satisfied. But it was clear to the FDA that essentially no dietary supplement product would meet the standard. Nevertheless, the widespread abuses of food labeling health claims that proliferated in the years immediately following the Kellogg's All-Bran ad were substantially reduced as a result of the passage of the NLEA.

The subsequent passage of the DSHEA in 1994 gave the dietary supplement industry significantly more freedom in the manufacturing and marketing of their products, including the types of claims that could be made on their labeling. Although these products were still effectively prevented from making explicit health/disease claims, the DSHEA did permit the use of structure/function claims on their labeling. And unlike the health claim provisions of the NLEA, the use of these structure/function claims did not require preapproval by the FDA. The dietary supplement industry was therefore able to quickly and easily incorporate these structure/function claims into the labels and advertising of a very wide assortment of products, and linking the products to a very wide spectrum of physiological functions. While the relationship between the supplement and a specific disease was not permitted under the DSHEA provisions, most consumers could easily make the connection from a structure/function claim to a specific disease or health condition. For example, it is not difficult to make the association between the structure/function claim related to "joint health" and the health/disease condition of "arthritis." Or the structure/function claim related to "mental function and memory" and the health/disease conditions of "Alzheimer's and dementia."

While the DSHEA resulted in the dissemination of perhaps hundreds of new structure/function claims on supplement labels, approval of new health claims based on the provisions of the NLEA was a slow and tedious regulatory process. Congress and the FDA attempted to expedite the approval process for health claims on conventional food labels by incorporating provisions into the FDA Modernization Act (FDAMA) of 1997 that would permit the use of health claims that were based on "authoritative statements" from federal scientific bodies. A federal scientific body would include *"a scientific body of the United States with official responsibility for public health protection or research directly related to human nutrition (such as the National Institutes of Health or the Center for Disease Control and Prevention) or the National Academy of Sciences (NAS) or any of its subdivisions."*[1] It is important to note, however, that FDAMA still upheld the "significant scientific agreement" standard when deciding whether to approve a health claim based on an "authoritative statement" from a federal scientific body. As noted in the FDA's guidance document on this topic[2]:

FDA intends to determine whether the standard of significant scientific agreement is met by a health claim based on an authoritative statement. And consistent with earlier regulations, FDA does not believe this standard would allow for a claim based on, for example, findings characterized as preliminary results, statements that indicate research is inconclusive, or statements intended to guide future research.

FDAMA does not provide for health claims based on authoritative statements for dietary supplements. This is because FDAMA amended the section of the Act that deals with procedures and standards for health claims for conventional foods, but did not amend the section that deals with procedures and standards for health claims for dietary supplements.

---

[1] 21 USC § 343(r)(2)(G)(i).
[2] *Guidance for Industry: Notification of a Health Claim or Nutrient Content Claim Based on an Authoritative Statement of a Scientific Body.* FDA. June 1998.

Thus, despite the FDAMA's intention to expedite the approval of health claims on food labeling, since its enactment in 1997, only six health claims based on authoritative statements have actually been approved.[3] In addition, the FDAMA had nothing to offer the dietary supplement industry in terms of expanded health claim labeling capabilities.

Still, many dietary supplement manufacturers wanted to go beyond the limitations associated with structure/function claims. Many of them argued that as long as a label claim did not include an objectively false statement, it should be considered First Amendment-protected free speech. Thus, for example, a label claim such as, "Some studies have shown that Vitamin C may reduce the incidence or severity of the common cold," is not technically a false statement. Of the hundreds of studies that have examined the relationship between vitamin C and the common cold, there were certainly a few that showed a slight beneficial effect. That is the nature of scientific discovery and research. There will always be some experimental results that disagree with the vast majority of other experimental results. The results of these few positive studies may have been due to random chance, or limitations in the experimental designs, or many other possible factors. To meet the "significant scientific agreement" standard expected of the FDA, the Vitamin C/common cold studies would need to pass the muster of peer review by qualified scientific experts and, most certainly, need to be reproducible over many repeat studies. Still, the original claim is not false. It is simply stating a fact; some studies have indeed shown positive effects of Vitamin C. Should a dietary supplement manufacturer be allowed to make such a "true" label claim, based on their free speech rights, even though the scientific consensus following careful review of all available data would clearly conclude that there is no benefit of Vitamin C on the common cold? This issue brings us to one of the most significant modifications of the FDA's enforcement of the NLEA and DSHEA since 1994.

For most of our U.S. history, commercial speech, which would include labeling and advertising of food products, was not considered a category of free speech protected under the First Amendment to the U.S. Constitution. The reasons for this may be somewhat obvious. The First Amendment essentially allows a citizen to say whatever they like, including falsehoods (as long as it is not under oath, or libelous, or promotes violence, or puts anyone in danger, such as falsely crying "fire" in a crowded theater). However, it could certainly be a problem for consumers if this right was extended to labeling and advertising. How would consumers know what to believe when deciding on purchases? In fact, the Food, Drug, and Cosmetic (FD&C) Act explicitly prohibits "labeling that is false or misleading in any

---

[3] *Guidance for Industry: A Food Labeling Guide (11. Appendix C: Health Claims).* FDA. January 2013.

particular."[4] So the government has historically been allowed to severely restrict commercial speech to protect consumers and allow for fair competition. However, beginning in the latter part of the 20th century, various court decisions began to grant more First Amendment protection to commercial speech. This culminated in a landmark U.S. Supreme Court case in 1980. In the Central Hudson Gas & Electric v. Public Service Commission case, the court ruled that it was a violation of the utility's First Amendment rights for the Public Service Commission to completely ban the utility's advertising to promote electricity use. As part of this ruling, the court applied a four-part test for determining when commercial speech violates the First Amendment. This test is still employed today for this purpose. Here is how it was stated in the Supreme Court ruling.[5]

> *In commercial speech cases, then, a four-part analysis has developed. At the outset, we must determine whether the expression is protected by the First Amendment. For commercial speech to come within that provision, [1] it at least must concern lawful activity and not be misleading. [2] Next, we ask whether the asserted governmental interest is substantial. If both inquiries yield positive answers, [3] we must determine whether the regulation directly advances the governmental interest asserted, and [4] whether it is not more extensive than is necessary to serve that interest.*

Thus, according to this test, *unlawful or misleading* commercial speech is not protected by the First Amendment. Second, in order for the government (in our case, the FDA) to attempt to regulate this commercial speech, it must establish that there is a substantial and important reason for the regulation. For example, to protect public health and safety or to avoid consumer confusion. Third, any regulation that the government imposes on the commercial speech must clearly address this important and specific concern that the government has identified. And fourth, any regulation that the government imposes on the commercial speech must be no more than the minimum that is necessary to address the concern.

So how does all of this relate to the regulation of dietary supplements? Recall that according to the NLEA, health claims (as opposed to structure/function claims) must be preapproved by the FDA. In addition, in order for a health claim to be approved by the FDA, it must meet the "significant scientific agreement" standard. In the late 1990s, dietary supplement marketers Durk Pearson and Sandy Shaw formally petitioned the FDA for

---

[4] 21 USC 343(a)(1).
[5] Central Hudson Gas & Electric v Public Service Commission. 447 U.S. 557 (1980).

approval of four new health claims for their dietary supplement labels. The four requested health claims were as follows:

1. Consumption of antioxidant vitamins may reduce the risk of certain kinds of cancer.
2. Consumption of fiber may reduce the risk of colorectal cancer.
3. Consumption of omega-3 fatty acids may reduce the risk of coronary heart disease.
4. Consumption of 0.8 mg of folic acid in a dietary supplement is more effective in reducing the risk of neural tube defects than a lower amount in foods in common form.

The FDA rejected all four proposed health claims, primarily based on its determination that the scientific evidence for these claims was inconclusive, and thus they did not meet the "significant scientific agreement" standard. Pearson and Shaw, as well as other plaintiffs with an interest in the outcome, challenged the FDA's decision in Federal District Court. The decision of the U.S. District Court, District of Columbia concluded the following[6]:

> *The FDA Final Rules did not violate the NLEA, APA, First Amendment or Fifth Amendment. The FDA carried out Congress' mandate when it adopted the "significant scientific agreement" standard for dietary supplements, and when it used that standard to review the four specific health claims denied in this case. The FDA provided adequate reasons for adopting the standard and its decision was neither arbitrary nor capricious. Additionally, its interpretation of the statute was correct and not contrary to law. The "significant scientific agreement" standard satisfied the Central Hudson test and therefore did not violate the First Amendment. Further, the four proposed health claims rejected in this case were inherently misleading, and therefore not protected by the First Amendment. Finally, the "significant scientific agreement" standard did not violate the Fifth Amendment because it sets out a specific and clear standard of review on a case-by-case review of the scientific evidence to determine whether there is significant scientific agreement among the experts that the proposed health claim is valid. For the reasons discussed above, Plaintiffs' Motion for Summary Judgment is denied and Defendants' Motion to Dismiss is granted. An Order will issue with this Opinion.*

[6] Pearson v. Shalala, 14 F. Supp. 2d 10 (D.D.C. 1998).

While this judicial decision appeared to be a clear victory for the FDA, the celebration was short-lived. Pearson and the other plaintiffs appealed the decision to the U.S. Court of Appeals for the District of Columbia.[7] As in the original District Court case, the FDA argued that the scientific data related to the four proposed health claims were "inconclusive" and thus did not meet the "significant scientific agreement" standard required of all health claims. As such, the claims were "inherently misleading" (the first of the four criteria of the Central Hudson Test), and therefore they were not protected as free speech under the First Amendment. The plaintiffs did not necessarily disagree that the scientific evidence in support of the claims may have been inclusive. But they argued that, even if the claims were "potentially misleading," this could be addressed without the need for the FDA to outright ban the claim. They contended that the inclusion of disclaimer language accompanying the claim would be an appropriate means of mitigating any potential consumer confusion while still protecting the plaintiff's commercial free speech rights. The FDA argued, however, that even if the claims were only "potentially misleading," it did not have to consider the use of disclaimers to mitigate any potential consumer confusion. So, the plaintiffs in this case were focusing their argument on the fourth condition of the Central Hudson Test. That is, that the FDA's "outright ban" of the four health claims was more extensive than necessary to serve the FDA's interest in protecting consumers from misleading claims. In addition, the plaintiffs argued that the "significant scientific agreement" standard required by the FDA for approval of any new health claim was in violation of the Administrative Procedure Act (APA) of 1946. The APA established rules under which federal agencies, such as the FDA, were permitted to establish and enforce new regulations. Congress passed this law to help ensure that Executive Branch enforcement agencies, such as the FDA, did not exceed or abuse their power and authority when writing regulations. In this case, the plaintiffs contended that the FDA did not clearly specify how it defined and applied the "significant scientific agreement" standard, and thus the application of this standard was "arbitrary and capricious" and in violation of the APA.

The U.S. Court of Appeals ruled in favor of the plaintiffs on both First Amendment grounds, as well as violation of the APA grounds. With regard to the First Amendment issues, the court rejected the FDA's contention that any health claim that failed to meet the "significant scientific agreement" standard was inherently misleading and therefore should be banned. To quote from the court's decision[8]:

---

[7] Pearson v. Shalala, 164 F. 3d 650 (D.C. Cir. 1999).
[8] ibid.

As best we understand the government, its first argument runs along the following lines: that health claims lacking "significant scientific agreement" are inherently misleading because they have such an awesome impact on consumers as to make it virtually impossible for them to exercise any judgment at the point of sale. It would be as if the consumers were asked to buy something while hypnotized, and therefore they are bound to be misled. We think this contention is almost frivolous. We reject it. But the government's alternative argument is more substantial. It is asserted that health claims on dietary supplements should be thought at least potentially misleading because the consumer would have difficulty in independently verifying these claims. We are told, in addition, that consumers might actually assume that the government has approved such claims.

Thus, the court essentially agreed with the FDA that consumers may find it difficult to interpret the truthfulness of potentially misleading claims. However, the court also recognized that disclaimers that accompany the claim may be enough to correct the potential consumer confusion. The disclaimer can be anything from, "The FDA has determined that the evidence supporting this claim is inconclusive," to an even more direct disclaimer such as, "The FDA does not approve this claim."

It is important to note, however, that the court's decision did not require that the FDA approve the four health claims originally proposed by Pearson and Shaw. Rather, it is simply required that the FDA considers the extent to which disclaimers may be employed to eliminate consumer confusion from a potentially misleading health claim. This is clearly conveyed in the following quote from the decision.[9]

We do not presume to draft precise disclaimers for each of appellants' four claims; we leave that task to the agency in the first instance. Nor do we rule out the possibility that where evidence in support of a claim is outweighed by evidence against the claim, the FDA could deem it incurable by a disclaimer and ban it outright. For example, if the weight of the evidence were against the hypothetical claim that "Consumption of Vitamin E reduces the risk of Alzheimer's disease," the agency might reasonably determine that adding a disclaimer such as "The FDA has determined that no evidence supports this claim" would not suffice to mitigate the claim's misleadingness.

[9] ibid.

The U.S. Court of Appeals also ruled on the appellants claim that the FDA violated the APA by not specifying how it interpreted and applied its "significant scientific agreement" standard when deciding whether to deny or approve a proposed health claim. The court's decision is explained as follows[10]:

> *...we agree with appellants that the APA requires the agency to explain why it rejects their proposed health claims – to do so adequately necessarily implies giving some definitional content to the phrase "significant scientific agreement." We think this proposition is squarely rooted in the prohibition under the APA that an agency not engage in arbitrary and capricious action. It simply will not do for a government agency to declare – without explanation – that a proposed course of private action is not approved. The agency must articulate a satisfactory explanation for its action. To refuse to define the criteria it is applying is equivalent to simply saying no without explanation. Accordingly, on remand, the FDA must explain what it means by significant scientific agreement or, at minimum, what it does not mean.*

In response to the U.S. Court of Appeals decision, the FDA embarked on a new strategy for dealing with proposed health claims that failed to meet the "significant scientific agreement" standard, but for which there was still some level of scientific evidence in support of the claim. As a result, a new category of food label health claims was created: *qualified health claims.* These health claims would need to include some type of qualifying "disclaimer" language to mitigate any potentially misleading aspects of the claim. According to the FDA, a food manufacturer (including both conventional foods and dietary supplements) who is interested in using a qualified health claim on their labeling would need to obtain premarket approval from the agency. This process would be similar to that which was already required for the premarket approval of authorized (unqualified) health claims. According to the FDA's "Guidance for Industry" document on this process, the manufacturer would formally petition the FDA for approval of the qualified health claim.[11] The FDA would then prioritize the petition based on various factors and would post the petition on its website for public comment. It would then seek either internal or

[10] ibid.
[11] Guidance for Industry: Interim Procedures for Qualified Health Claims in the Labeling of Conventional Human Food and Human Dietary Supplements. U.S. Food and Drug Administration. July 2003.

external/third-party (expert advisory committee) scientific review of the data submitted in support of the claim. Based on the recommendations from the scientific review and public comments, the FDA may decide to issue a "letter of enforcement discretion." This is defined by the FDA as follows[12]:

> *A letter of enforcement discretion is a letter issued by FDA to the petitioner specifying the nature of the qualified health claim for which FDA intends to consider the exercise of its enforcement discretion. If a letter of enforcement discretion has been issued, FDA does not intend to object to the use of the claim specified in the letter, provided that the products that bear the claim are consistent with the stated criteria.*

In effect, the FDA is using its "enforcement discretion" to either approve or reject a proposed qualified health claim, without going through a formal rulemaking process. In determining the type of disclaimer language that may be required for an approved qualified health claim, the FDA originally established criteria based on a "scientific ranking" of the evidence in support of the claim. An adapted version of this ranking system is illustrated in the following table.[13]

Scientific ranking	FDA category	Appropriate qualifying language
First level	A	*Meets "significant scientific agreement" standard. No qualifying language is necessary.*
Second level	B	*…"although there is scientific evidence supporting the claim, the evidence is inconclusive."*
Third level	C	*"Some scientific evidence suggests… however, FDA has determined that this evidence is limited and not conclusive."*
Fourth level	D	*"Very limited and preliminary scientific research suggests… FDA concludes that there is little scientific evidence supporting the claim."*

[12] Guidance for Industry: FDA's Implementation of "Qualified Health Claims": Questions and Answers: Final Guidance. U.S. Food and Drug Administration. May 12, 2006.
[13] Consumer Health Information for Better Nutrition Initiative" Task Force Final Report. U.S. Food and Drug Administration. July 10. 2003.

Thus, an FDA "A" category health claim was equivalent to an authorized "unqualified" health claim. The remaining three FDA categories represented increasingly weak scientific evidence in support of the claim. Since it began applying its qualified health claim petition and review process in 2003, the FDA has issued letters of enforcement discretion in response to 33 petitions.[14] Many of these petitions involved multiple proposed qualified health claims related to a particular food or ingredient. For example, in May of 2004, the H. J. Heinz Company petitioned the FDA for approval of the following qualified health claims[15]:

1) "Although the evidence is not conclusive, tomato lycopene may reduce the risk of prostate cancer."

2) "Although the evidence is not conclusive, tomato lycopene may reduce the risk of prostate cancer when consumed as part of a healthy diet."

3) "Although the evidence is not conclusive, tomato products, which contain lycopene, may reduce the risk of prostate cancer."

4) "Although the evidence is not conclusive, tomatoes and tomato products, which contain lycopene, may reduce the risk of prostate cancer."

5) "Although the evidence is not conclusive, tomato products, which contain lycopene, may reduce the risk of prostate cancer when consumed as part of a healthy diet."

6) "Although the evidence is not conclusive, tomatoes and tomato products, which contain lycopene, may reduce the risk of prostate cancer when consumed as part of a healthy diet."

7) "Although the evidence is not conclusive, lycopene in tomato products may reduce the risk of prostate cancer."

8) "Although the evidence is not conclusive, lycopene in tomatoes and tomato products may reduce the risk of prostate cancer."

9) "Although the evidence is not conclusive, lycopene in tomato products may reduce the risk of prostate cancer when consumed as part of a healthy diet."

[14] Qualified Health Claims: Letters of Enforcement Discretion. U.S. Food and Drug Administration.www.fda.gov/Food/IngredientsPackagingLabeling/LabelingNutrition/ucm072756.htm.
[15] *Qualified Health Claims: Letter Regarding Tomatoes and Prostate Cancer (Lycopene Health Claim Coalition)* (Docket No. 2004Q-0201). U.S. Food and Drug Administration. November 8, 2004.

10) "Although the evidence is not conclusive, lycopene in tomatoes and tomato products may reduce the risk of prostate cancer when consumed as part of a healthy diet."

11) "Although the evidence is not conclusive, lycopene in fruits and vegetables, including tomatoes and tomato products, may reduce the risk of prostate cancer."

On November 8, 2005, the FDA issued its final decision regarding this petition.[16] In the agency's detailed response, it thoroughly and systematically reviewed the scientific evidence related to the claims. The FDA's conclusion was as follows:

Based on FDA's consideration of the scientific evidence submitted with your petition, and other pertinent scientific evidence, FDA concludes that there is no credible evidence to support a qualified health claim for tomato lycopene; tomatoes and tomato products, which contain lycopene; lycopene in tomatoes and tomato products; lycopene in fruits and vegetables, including tomatoes and tomato products, and lycopene as a food ingredient, a component of food, or as a dietary supplement and reduced risk of prostate cancer. Thus, FDA is denying these claims. However, FDA concludes that there is very limited credible evidence for qualified health claims for tomatoes and/or tomato sauce, and prostate cancer provided that the qualified claim is appropriately worded so as to not mislead consumers. Thus, FDA intends to consider exercising its enforcement discretion for the following qualified health claim:

Prostate Cancer

Very limited and preliminary scientific research suggests that eating one-half to one cup of tomatoes and/or tomato sauce a week may reduce the risk of prostate cancer. FDA concludes that there is little scientific evidence supporting this claim.

Scanning through any of the 33 petitions, it becomes quickly clear that the FDA is still holding manufacturers (both for conventional foods and dietary supplements) to a strict evaluation and interpretation of the

---

[16] ibid.

available scientific data related to the proposed qualified health claims. Most of the proposed health claims are denied outright based on the FDA's conclusion that "there is no credible evidence to support a qualified health claim." Of those claims that are approved, they are often required to include the most strict disclaimer language (what would be considered close to the FDA's former Category "D" scientific ranking). It is perhaps not surprising, therefore, that despite nearly two decades since the Pearson v. Shalala Appeals Court decision, qualified health claims have failed to become widely popular on conventional food and dietary supplement labeling. While there are no extensive data available on the extent to which qualified health claims are currently used in food and supplement labeling, one study from 2009 reported that structure–function claims are used much more frequently, particularly by dietary supplement manufacturers.[17] These investigators further speculated on why unqualified health claims have not been used more often. Among some of the noted reasons are the fact that structure–function claims, unlike qualified health claims, do not have to be preapproved by the FDA, saving manufacturers enormous amounts of time and money preparing their labeling and marketing. In addition, structure–function claims require minimal scientific proof, as long as the claims are not false or misleading. Finally, a study conducted by the International Food Information Council in 2005 found that consumers rate the scientific evidence in support of qualified health claims similar to how they rate the scientific evidence in support of a structure–function claim.[18] In addition, consumers are more likely to purchase products labeled with a structure–function claim over products with FDA's B, C, or D level disclaimers for qualified health claims. So, there is apparently little incentive or advantage for supplement manufacturers to use qualified health claims over structure–function claims.

The Pearson v. Shalala decision also required that the FDA establishes and makes public its policy and procedures for determining whether a claim meets its "significant scientific agreement" standard. In January 2009, the FDA published its most recent "Guidance for Industry" on this topic.[19] Below is how the agency describes its definition of "significant scientific agreement" in this report.

---

[17] Bone, PF and KR France. Qualified Health Claims on Package Labels. *J. Public Policy Mark.*, 28(2): 253–258 (2009).

[18] International Food Information Council. *2004 Qualified Health Claims Research Executive Summary.* www.foodinsight.org/2004_Qualified_Health_Claims_Research_Executive_ Summary (Accessed Sept. 2017).

[19] *Guidance for Industry: Evidence-Based Review System for the Scientific Evaluation of Health Claims – Final.* U.S. Food and Drug Administration. January 2009.

FDA's determination of SSA [significant scientific agreement] represents the agency's best judgment as to whether qualified experts would likely agree that the scientific evidence supports the substance/disease relationship that is the subject of a proposed health claim. The SSA standard is intended to be a strong standard that provides a high level of confidence in the validity of the substance/disease relationship. SSA means that the validity of the relationship is not likely to be reversed by new and evolving science, although the exact nature of the relationship may need to be refined. SSA does not require a consensus based on unanimous and incontrovertible scientific opinion. SSA occurs well after the stage of emerging science, where data and information permit an inference, but before the point of unanimous agreement within the relevant scientific community that the inference is valid.

In this document, the agency also details the types of scientific studies it would expect to be included in its evaluation of the proposed claims, how it will assess the quality of the studies, and how it will evaluate the totality of the scientific evidence. This evidence-based review system would be employed to make decisions regarding both authorized (unqualified) health claim petitions (those that meet the significant scientific agreement standard), as well as qualified health claim petitions. With the publication of this 2009 Guidance for Industry, the FDA has indicated that it will no longer utilize the "A, B, C, D" ranking system that it proposed in 2003. However, it may reconsider a ranking system in the future, depending on the outcome of ongoing studies designed to evaluate consumer understanding "of ranking systems that could be used to describe the strength of scientific evidence for a health claim."[20]

It is clear that the FDA, the dietary supplement industry, and the consuming public have come a long way since the All-Bran cereal ads and labeling with regard to health claims. Nevertheless, as a result of the combination of legislative actions, court challenges, and ever-evolving FDA enforcement strategies, there are now several ways of communicating health information to the public on food and dietary supplement labeling, each with its own regulatory framework. Here is a summary of these labeling options.

Nutrient Content Claims: These types of labeling claims simply communicate the relative level of a particular nutrient or other nutritional characteristic of the food. Examples of these types of claims would be, "low fat," "high fiber," or "contain 100 calories." The NLEA of 1990 set forth a legal framework for the types of nutrient content claims permitted on

---

[20] ibid.

food labeling. The FDA has promulgated detailed regulations enforcing all aspects of these types of claims. Nutrient content claims are sometimes referred to as "implicit" or "implied" health claims. This is because they do not make any direct connection to a particular disease or health condition. Rather, they rely on the consumer to make the connection. For many of these claims, the connection is rather obvious to the consumer. For example, most consumers would likely make the correct connection between a nutrient content claim of "low sodium" and the possible health benefits related to hypertension. Nutrient content claims do not require preapproval by the FDA. Manufacturers simply need to follow the appropriate labeling regulations.

Dietary Guidance Statements: According to the FDA, dietary guidance statements "make reference to a category of foods (i.e., a grouping that is not readily characterized compositionally) and not to a specific substance." An example of these types of claims would be, "Diets low in saturated fat and cholesterol may reduce your risk of heart disease." Like nutrient content claims, dietary guidance statements do not require premarket approval by the FDA. Food manufacturers simply need to ensure that the statements are truthful and not misleading.

Authorized Health Claims: Authorized health claims are explicit disease-related health claims that meet FDA's "significant scientific agreement" standard and which have been approved by the FDA following submission of a formal petition by the manufacturer. Currently, there are only 12 approved "authorized health claims." An example of one of these label claims would be, "Development of cancer depends on many factors. A diet low in total fat may reduce the risk of some cancers."[21]

Health Claims Based on Authoritative Statements: As discussed earlier in this chapter, these types of explicit health claims are permitted based on the provisions of the FDAMA of 1997. As with authorized health claims, these claims need to be preapproved by the FDA before being used by food manufacturers. In order to be approved, the claim must still meet the "significant scientific agreement" standard and must be based on an "authoritative statement" from certain scientific bodies of the U.S. Government, such as the United States Department of Agriculture (USDA), the National Academy of Sciences, or the National Institutes of Health. Also as mentioned earlier, the provisions of the FDAMA do not allow for the use of these types of claims on the labeling of dietary supplements. Since the passage of the FDAMA in 1997, the FDA has only approved six health claims based on authoritative statements. An example of one of these is, "Diets containing foods that are a good source of

---

[21] *Guidance for Industry: A Food Labeling Guide (11. Appendix C: Health Claims).* U.S. Food and Drug Administration. January 2013.

potassium and that are low in sodium may reduce the risk of high blood pressure and stroke."[22]

Qualified Health Claims: These health claims may be permitted if they are supported by some level of scientific evidence, but fail to meet the "significant scientific agreement" standard required for authorized (unqualified) health claims. To minimize any potential confusion on the part of consumers due to the varying degrees of scientific evidence in support of the claim, the FDA will require that some disclaimer language be included with the health claim. Food manufacturers must formally petition the FDA for approval of a new qualified health claim, before using the claim on their product labeling. An example of an approved qualified health claim is, "Some scientific evidence suggests that consumption of antioxidant vitamins may reduce the risk of certain forms of cancer. However, the FDA does not endorse this claim because the evidence is limited and not conclusive."[23]

Structure–Function Claims: These claims are permitted based on the provisions of the DSHEA of 1994. Structure–function claims simply describe the effect a substance (nutrient or ingredient) may have on the structure or function of some part of the body. For example, "Calcium builds strong bones," or "Vitamin A helps maintain a healthy immune system." The most important distinction between a "structure–function claim" and an "authorized" or "qualified" health claim is that structure–function claims do not refer to a specific disease. Another important distinction is that structure–function claims do not require premarket approval by the FDA. The manufacturer simply needs to be able to substantiate that the claim is truthful and not misleading (if requested), and they must notify the FDA with the specific text of the claim within 30 days of marketing the product with the claim. In addition, the label must bear the required structure–function claim disclaimer, which reads, "This statement has not been evaluated by the Food and Drug Administration. This product is not intended to diagnose, cure, treat, or prevent any disease."

We are now almost up to date with the current status of dietary supplement regulation. In the final chapter of this book, we will review some of the more recent legislative actions that have impacted the regulation of dietary supplements. We will conclude with a discussion some of the strengths, weaknesses, and challenges of the current regulatory framework and some of the attempts that have been made to improve the regulation of dietary supplement in the United States.

---

[22] ibid.
[23] Summary of Qualified Health Claims Subject to Enforcement Discretion. U.S. Food and Drug Administration. www.fda.gov/food/ingredientspackaginglabeling/labelingnutrition/ucm073992.htm.

chapter fifteen

Where does dietary supplement regulation stand today?

For nearly 25 years now, the Dietary Supplement Health and Education Act (DSHEA) has provided the dietary supplement industry with wide-ranging statutory protections from Food and Drug Administration (FDA) regulatory intrusion. Critics of the DSHEA are quick to point out what they see as many of the weaknesses of this law, in terms of protecting the consuming public from potentially harmful, ineffective, or mislabeled products. It is worth noting, however, that there have been some important legislative actions that have impacted the regulation of dietary supplements over the past two decades since the passage of the DSHEA. For example, in 2006 President George W. Bush signed into law the Dietary Supplement and Nonprescription Drug Consumer Protection Act. This law requires that a "manufacturer, packer, or distributor" of a dietary supplement report to the FDA any serious adverse events received regarding a particular dietary supplement. The law defines an "adverse event" and a "serious adverse event" as follows:

---

21 U.S. code (USC) § 379aa-1

The term "adverse event" means any health-related event associated with the use of a dietary supplement that is adverse.

The term "serious adverse event" is an adverse event that—
(A) results in—
    (i)   death;
    (ii) a life-threatening experience;
    (iii) inpatient hospitalization;
    (iv) a persistent or significant disability or incapacity; or
    (v) a congenital anomaly or birth defect;
    or
(B) requires, based on reasonable medical judgment, a medical or surgical intervention to prevent an outcome described under subparagraph (A).

According to this law, the responsible person is only required to report "serious adverse events" to the FDA. However, they are still required to keep records of all adverse events received, including non-serious events. The FDA has the authority to review these records as part of its normal inspection procedures.

Not all legislative attempts to strengthen dietary supplement regulation were successful. In 2010, Senator John McCain from Arizona introduced the Dietary Supplement Safety Act into Congress.[1] The bill was intended to address some of the perceived weaknesses in the DSHEA of 1994. Specifically, it would have required dietary supplement manufacturers to register their facility with the FDA. It also would have required manufacturers of dietary supplements containing new dietary ingredients (NDI's) to provide evidence establishing the safety of the new ingredient. In addition, manufacturers would be required to collect and report both serious and non-serious adverse events to the FDA. And finally, it would have given the FDA mandatory recall authority for any dietary supplement that was determined to be a hazard to health. However, less than 1 month after introducing the bill, Senator McCain withdrew his support after meeting with Senator Hatch, who you may recall from an earlier chapter of this book, is a very strong advocate for the dietary supplement industry. Below is a copy of the letter from Senator Hatch to Senator McCain acknowledging Senator McCain's decision to withdraw his support for the legislation.[2]

It is unclear how Senator Hatch was able to persuade Senator McCain to withdraw his support for the bill. Perhaps it was partly due to other legislation that was working its way through the Congress around that time. This new legislation would ultimately address at least some of the concerns that Senator McCain had regarding the regulation of dietary supplements. On January 4, 2011, President Obama signed into law the Food Safety Modernization Act (FSMA). This was a comprehensive and far-reaching amendment to the Food, Drug, and Cosmetic (FD&C) Act. Dietary supplements were not the principle focus of FSMA. Rather, it dealt primarily with food sanitation, regulation of imports, and many other aspects related to broad issues of food safety. However, some important provisions of the law did address issues that were particularly relevant to dietary supplements. First, it required all food manufacturers (with limited exceptions) to register their facilities with the FDA (similar to what was proposed in Senator McCain's bill). Since dietary supplements are, by legal definition, a category of "food," these manufacturers were now

---

[1] S. 3002 — 111th Congress: Dietary Supplement Safety Act of 2010." www.GovTrack.us. 2010. September 30, 2017 <www.govtrack.us/congress/bills/111/s3002>

[2] http://fdcalerts.typepad.com/files/100308_hatch_mccain_s3002_letter.pdf.

UNITED STATES SENATE
WASHINGTON, D. C.

ORRIN G. HATCH
Utah

March 4, 2010

The Honorable John McCain
241 Russell Senate Office Building
Washington, DC 20510

Dear John:

Thank you for taking the time to talk with me this afternoon about S. 3002, the Dietary Supplement Safety Act of 2010.

I am pleased that you understand my serious concerns with your bill. Also, I want to thank you for agreeing to withdraw your support for the provisions of S. 3002 that I believe would do great harm to the dietary supplement industry and work with me on solutions that will truly help dietary supplement consumers without injuring this important industry. More than 100 million Americans regularly consume dietary supplements as a means of improving and maintaining healthy lifestyles. Therefore, continued access to these products is extremely important to them.

As you know, I authored the 1994 Dietary Supplement Health and Education Act (DSHEA) which ensures that individuals would have access to safe supplements and information about their use. DSHEA provided the Food and Drug Administration (FDA) with all the authority they need, and more authority than they had ever had previously, to remove harmful substances from store shelves and ensure consumers have access to safe supplements.

As we move forward on this important issue, I want to work with you on calling for the full enforcement of existing laws, such as DSHEA, so Americans will have uninterrupted access to safe dietary supplements and bad actor companies are removed from the market immediately.

I'm counting on you working with me to make sure this important industry does not fall prey to over-regulatory regimes and mounds of costly government bureaucracy. I believe that if you and I will work together, we will be able to make a difference for good in the dietary supplement industry. In short, we will ensure safe supplements are always available to consumers and the dietary supplement industry will be able to continue to innovate and grow.

John, thanks for your willingness to resolve this important issue.

Sincerely,

Orrin G. Hatch
United States Senator

subject to this requirement. In addition, FSMA gave the FDA mandatory recall authority over all foods, which once again would include dietary supplements. So two of the proposed provisions of Senator McCain's earlier Dietary Supplement Safety Act were now covered in FSMA.

Other provisions of FSMA are also having a significant impact on the dietary supplement industry. The principal goal of FSMA was to institute a proactive federal regulatory framework that would prevent food safety problems before they occur, rather than deal with the consequences after they occur. Thus, a significant proportion of FSMA's provisions and the FDA's regulations enforcing the provisions focus on identifying potential hazards and establishing preventive controls to avoid the hazards. But recall also, from Chapter thirteen, that the FDA had already finalized Current Good Manufacturing Practice (CGMP) regulations specific for dietary supplement manufactures in 2007. So many of the *hazard analysis* and *preventive control* requirements of FSMA were already covered for dietary supplements in the CGMP regulations. In fact, one section of FSMA addresses this issue.

> **21 USC §350(g) Notes:** Nothing in the amendments made by this shall apply to any facility with regard to the manufacturing, processing, packing, or holding of a dietary supplement that is in compliance with the requirements of sections 402(g)(2) and 761 of the Federal FD&C Act (21 USC 342(g)(2), 379aa-1).

In other words, as long as the dietary supplement manufacturer is in compliance with the CGMP's for dietary supplements, the manufacturer is exempt from some, but not all, of the related FSMA regulations. However, most dietary supplements are manufactured with ingredients supplied from other companies, many of which may be located in other countries. These dietary supplement ingredient suppliers have no such exemption and must be in full compliance with FSMA laws and regulations. Of particular relevance, FSMA requires that dietary supplement ingredient manufacturers be in compliance with the provisions of the law pertaining to:

Prevention Control for Human Food: Manufacturers must identify possible hazards that may occur in the production of the ingredients, what prevention/control steps will be established to prevent these hazards, and what actions will be taken if problems arise.

Foreign Supplier Verification for Importers: According to the FDA, this is *"a program that importers covered by the rule must have in place to verify that their foreign suppliers are producing food in a manner that provides the same level of public health protection as the preventive controls or produce safety regulations, as appropriate, and to ensure that the supplier's food is not adulterated and is not misbranded with respect to allergen labeling."* Details on how companies can meet the requirements of FDA's final rule for this program can be found on the FDA's website.[3]

---

[3] US Food and Drug Administration. *FSMA Final Rule on Foreign Supplier Verification Programs (FSVP) for Importers of Food for Humans and Animals.* www.fda.gov/Food/GuidanceRegulation/FSMA/ucm361902.htm.

Accredited Third-Party Certification: This is a voluntary program for accrediting third-party certification "auditors," whose responsibility is to ensure that ingredient suppliers in other countries are in compliance with the U.S. laws. More information on FDA's implementation of this final rule can be found on the FDA website.[4]

Despite some of the improvements in dietary supplement regulation provided by the Dietary Supplement and Nonprescription Drug Consumer Protection Act in 2006 and FSMA in 2011, some members of Congress continued to fight for still more aggressive dietary supplement regulation. In June of 2011, Senator Richard Durbin (D-Illinois) introduced the Dietary Supplement Labeling Act.[5] The bill died in that Congress, but Senator Durbin reintroduced a nearly identical bill in August of 2013.[6] This bill went beyond the facility registration requirements of FSMA and additionally required that dietary supplement manufacturers provide the FDA with details on the products they produce, including all ingredients used in the products. The FDA would then make a determination as to the potential risk of "serious adverse events, drug interactions, or contraindications" of any of the ingredients. If such a risk were identified, particularly for high-risk subgroups of the population (such as children or pregnant or breastfeeding women), the FDA could require warning labels on the product. In addition, the bill gave the FDA the authority to require that manufacturers provide substantiation for any structure–function claims they include on their product labeling. Finally, the bill included the provision shown as follows.

> *The Secretary of Health and Human Services, not later than 1 year after the date of enactment of this Act and after providing for public notice and comment, shall establish a definition for the term conventional food for purposes of the Federal Food, Drug, and Cosmetic Act (21 U.S.C. 301 et seq.). Such definition shall take into account conventional foods marketed as dietary supplements, including products marketed as dietary supplements that simulate conventional foods.*

Once again, however, no vote was ever taken on this bill and it died in Congress. But Senator Durbin's bill did bring some necessary attention

[4] US Food and Drug Administration. FSMA Final Rule on Accredited Third-Party Certification. www.fda.gov/Food/GuidanceRegulation/FSMA/ucm361903.htm.

[5] S. 1310 — 112th Congress: Dietary Supplement Labeling Act of 2011." www.GovTrack.us. 2011. September 30, 2017 <www.govtrack.us/congress/bills/112/s1310.

[6] S. 1425 — 113th Congress: Dietary Supplement Labeling Act of 2013." www.GovTrack.us. 2013. September 30, 2017 <www.govtrack.us/congress/bills/113/s1425>

to one of the more controversial interpretations of the DSHEA. That is, declaring a food to be a "dietary supplement" in order to take advantage of the DSHEA's weak safety requirements for dietary supplement ingredients. This issue drew some national attention in 2011 when a product known as "Lazy Cakes Brownies" hit the market. Melatonin was an active ingredient in these brownie snack cakes. Melatonin is a neurohormone produced naturally in the pineal gland of the brain. It plays a role in regulating the sleep–wake cycle. It is also a popular dietary supplement ingredient, sold in pill form, purportedly to help with jet lag and other sleep issues. The labeling of the "Lazy Cakes Brownies" packaging clearly declared the product to be a "DIETARY SUPPLEMENT," and it included the standard dietary supplement disclaimer at the bottom. The back of the package even had the required "Supplement Facts" panel. The product was marketed as a "relaxation" dietary supplement.

So, what did the FDA find problematic with this product? You may recall from an earlier chapter of this book that the FDA had once considered regulating ingredients in dietary supplements as "food additives." According to the FD&C Act, this would have required manufacturers to prove that new food additives were "safe" before they would be allowed for use in foods. Dietary supplement manufacturers were understandably not pleased with this proposed regulatory strategy. In fact, it was the FDA's consideration of this regulatory approach that was a major driving force behind the passage of the DSHEA. The DSHEA amended the legal definition of a food additive to specifically exclude *"ingredients....intended for use in a dietary supplement."*[7] Thus, according to the DSHEA, if a manufacturer of a product wants to avoid having to conduct very expensive and time-consuming safety testing of a potential "food additive" ingredient in their product, they can simply declare the product to be a "dietary supplement." This may very well have been the intention of the manufacturers of "Lazy Cakes Brownie" product. But does this product truly meet the legal definition of a dietary supplement? Back in Chapter ten, we discussed the major provisions of the DSHEA, including its broad definition of the term "dietary supplement." However, this section of the DSHEA also includes an important statement that is particularly relevant to the "Lazy Cakes" product.

[21 USC § 321(ff)]
The term "dietary supplement"—
(2) means a product that—
    (B) is not intended for use as a conventional food or as a sole item of a meal or the diet;

[7] 21 USC §321(s)(6).

In August of 2011, the FDA concluded that "Lazy Cakes" brownies were "intended for use as a conventional food" (a brownie) and thus were excluded from the legal definition of a dietary supplement. As a conventional food product (rather than a dietary supplement), the FDA was acting appropriately in declaring that this product contained an unapproved food additive (melatonin) and was therefore adulterated according to 21 USC §342(a)(2)(C). This case also points out some of the paradoxes in the regulation of conventional foods versus dietary supplements. On the one hand, melatonin when used as an ingredient in a conventional food (Lazy Cakes) is an unapproved food additive (requiring premarket safety testing), while melatonin in potentially much higher doses in a dietary supplement pill is permitted by the DSHEA without any required premarket safety testing.

Since the passage of the DSHEA in 1994, the FDA has often had to deal with products that blurred the distinction between dietary supplements and conventional foods. This is particularly true of many liquid or beverage-like products, such as energy and caffeinated drinks. In January of 2014, the FDA published its "Guidance for Industry" document on this topic.[8] In deciding whether a product should be regulated as a liquid dietary supplement versus a conventional beverage food, the FDA considers a number of factors, including how the product is labeled and advertised, the name of the product, the packaging design, the serving size, how it is marketed, and its ingredients. If the FDA determines that the product meets the definition of a conventional food, then it will enforce the appropriate regulations pertaining to the labeling and formulation (ingredients) for conventional foods.

Where does dietary supplement regulation go from here?

Now that we are essentially up-to-date with our review of the current state of dietary supplement regulation, it may be appropriate to revisit a question raised in the very first chapter of this book. Are the DSHEA and related laws (the Dietary Supplement and Nonprescription Drug Consumer Protection Act of 2006 and the FSMA of 2011), providing the FDA with sufficient regulatory authority to adequately protect the public from potentially dangerous dietary supplement products? Or from products that make misleading or fraudulent claims? The answers to those questions will certainly depend on whom you ask. There is little doubt

---

[8] Guidance for Industry: Distinguishing Liquid Dietary Supplements from Beverages. U.S. Food and Drug Administration. January 2014. www.fda.gov/food/guidanceregulation/guidancedocumentsregulatoryinformation/ucm381189.htm.

that the vast majority of dietary supplements on the market today are relatively safe, at least with regard to so-called "direct" toxicity risks. The indirect risks mentioned in earlier chapters (those that are related to the reliance on dietary supplements to treat ailments that may be more effectively treated with mainstream medical approaches), are much more difficult to estimate. Nevertheless, the dietary supplement industry will often compare the safety record of dietary supplements to that of prescription drugs as proof that the vast majority of dietary supplements products are very safe and adequately regulated. Even many over-the-counter (OTC) drugs present risks that are much higher than most dietary supplements. For example, a recent study on the safety of acetaminophen (the active ingredient in OTC drugs like Tylenol) reported that there were approximately 59,000 emergency department visits per year between 2004 and 2012 related to adverse drug events.[9] And that is for just one OTC product category. By comparison, a recent study conducted by scientists from the Centers for Disease Control and Prevention and the FDA estimated that there were approximately 23,000 visits per year to hospital emergency departments due to adverse effects of dietary supplements.[10] And that is a figure for a product category that represents thousands of individual products. This estimate included injuries ranging from allergic reactions, heart trouble, nausea and vomiting, and choking (often due to large pill sizes and associated swallowing problems in older users). Weight loss and energy products were responsible for many of the heart problem issues, particularly in younger users. And approximately 10% of the cases were serious enough to require hospitalization. Of course, these estimates likely reflect only those adverse events that were acute in nature, causing the individual to seek immediate medical attention. It is much more difficult to estimate how many additional chronic serious adverse events are occurring due to long-term consumption of particular dietary supplements.

With the passage of the Dietary Supplement and Nonprescription Drug Consumer Protection Act in 2006, the FDA had hoped to be better able to identify dangerous dietary supplements through this law's "serious adverse event" reporting system requirement. But the "serious adverse event" approach to protecting the public from dangerous dietary supplements is akin to "closing the barn door after the horse has bolted." We have already experienced the consequences of finding out too late that

---

[9] Major, JM, EH Zhou, HL Wong, JP Trinidad, TM Pham, H Mehta, Y Ding, JA Staffa, S Iyasu, C Want, ME Willy. Trends in Rates of Acetaminophen-related adverse events in the United States. *Pharmacoepidemiol. Drug Saf.*, 25: 590–598 (2016).

[10] Geller, AI, N Shehab, NJ Weidle, MC Lovegrove, BJ Wolpert, BB Timbo, RP Mozersky, DS Budnitz. Emergency Department Visits for Adverse Events Related to Dietary Supplements. *N. Engl. J. Med.*, 373(16): 1531–1339 (2015).

a dietary supplement was dangerous. Consider the 1,500 illnesses and 38 deaths associated with the L-tryptophan Eosinophilia Myalgia Syndrome event of the late 1980s (discussed in Chapter eight). Or the approximately 16,000 adverse event reports for products containing the herbal stimulant ephedra, marketed as a weight loss and athletic performance aid. This dietary supplement ingredient had been linked to as many as 155 deaths, including that of Baltimore Orioles pitcher Steve Bechler in 2003.[11] The FDA finally banned ephedra-containing dietary supplements in 2004. More recently, a 2013 outbreak of acute hepatitis and liver failure in Hawaii was linked to the dietary supplement, OxyElitePro, a product sold as a weight loss and muscle building supplement.[12] This product was linked to several cases of liver failure (requiring liver transplantation) and at least one death. In the fall of 2013, the FDA used the threat of its FSMA mandatory recall authority to convince the manufacturers of this product to conduct a voluntary recall.

Liver toxicity (hepatotoxicity) is now considered to be one of the most common "serious adverse events" linked to some dietary supplements. From a physiological and biochemical perspective, this makes quite a bit of intuitive sense. Among the liver's main functions is to receive and process blood that is carrying absorbed substances from the intestinal tract (such as ingested dietary supplement chemicals). It is also a principle site for the metabolism of drugs and detoxification of many chemicals (including ingested dietary supplements). Recent studies have suggested that as much as 20% of liver injury cases in the United States may be attributed to the use of dietary supplements.[13] This is a staggering health statistic and reflects a large increase over the previous decades, likely due to the increasing use and variety of dietary supplements over this period. The types of dietary supplement products linked to liver injury vary widely. Often they will contain multiple ingredients that may include synthetic derivatives of metabolically active chemicals, various herbal and botanical ingredients, and less frequently, simple vitamins or minerals.

It is precisely this highly varied nature of the formulations of many dietary supplement products that make them so difficult to test for safety and effectiveness. Consider that before the passage of the DSHEA in 1994, there were estimated to be approximately 4,000 different dietary

---

[11] *Why the FDA Banned Ephedra*. Harvard Health Publishing. Harvard Medical School. March 2004. (www.health.harvard.edu/staying-healthy/ephedra-ban).

[12] Johnston, DI, A Chang, M Viray, K Chatham-Stephens, H He, E Taylor, LL Wong, J Schier, C Martin, D Fabricant, M Salter, L Lewis, SY Park. Hepatotoxicity Associated With the Dietary Supplement OxyELITE Pro – Hawaii, 2013. *Drug Test. Anal.*, 8(3–4): 319–327 (2016).

[13] Navarro, VJ, I Khan, E Bjornsson, LB Seeff, J Serrano, JH Hoofnagle. Liver Injury from Herbal and Dietary Supplements. *Hepatology*, 65(1): 363–373 (2017).

supplement products on the market.[14] Today, there are well in excess of 55,000 different supplements in the marketplace. The inclusion of herbal and other botanical ingredients and products within the DSHEA's legal definition of what can be considered a "dietary supplement" created some unique and important new safety concerns. Before the passage of this law in 1994, most consumers probably thought of dietary supplements as primarily consisting of vitamin and mineral products. That is, products that provided the body with *nutrients*. There is a very large body of good scientific information on how our bodies absorb, metabolize, and excrete these substances. There is also a large body of good scientific data on the roles that vitamins and minerals play in health and disease, as well as the risks posed by consuming too much of them. Thus, even though the DSHEA (and the Vitamin and Mineral Amendment of 1976) did not permit the FDA to establish upper limits on the potency of vitamin or mineral supplements, at least the agency and other health professionals could offer consumers scientifically sound advice on how much of these nutrients are needed in the diet and how much could be safely consumed. In fact, the Food and Nutrition Board (FNB) of the National Academy of Medicine (within the National Academy of Sciences, Engineering, and Medicine) regularly publishes Dietary Reference Intakes (DRIs) for vitamins, minerals, and other nutrients. These DRIs are based on the latest scientific information related to the amounts of these nutrients needed to maintain health as well as the upper limits of safe intake. The DRIs consist of Recommended Dietary Allowances (RDAs) for nutrients for which the FNB believes there is good scientific agreement on which to base a daily recommendation of intake. In addition, the DRIs include "Adequate Intake" (AI) values for nutrients, for which the FNB believes there is not yet sufficiently strong scientific evidence on which to base an RDA recommendation. And finally, the DRI's include Tolerable Upper Intake Levels (ULs) for vitamins and minerals that represent the maximum amount of a nutrient that can be safely consumed on a regular basis. Now consider how many of the DRIs exist, relative to the number and variety of dietary supplements on the market today.

RDAs: Established for 10 vitamins, 9 minerals, total carbohydrate, and total protein.

AIs: Established for 4 vitamins, 6 minerals, total fiber, total fat, and 2 essential fatty acids.

Tolerable ULs: Established for 8 vitamins and 16 minerals.

---

[14] Cohen, PA. The Supplement Paradox. Negligible Benefits, Robust Consumption. *J. Am. Med. Assoc.*, 316(14): 1453–1454 (2016).

Thus, the top scientific authorities within the federal government have only established scientifically based recommendations (RDAs or AIs) for 14 vitamins, 15 minerals, and a few additional nutrients. And they have only identified safe ULs for 8 vitamins and 16 minerals. There are no similar recommendations for the vast majority of other chemicals that are in most dietary supplement products on the market today.

This lack of scientific information (related to both safety and effectiveness) for the vast majority of dietary supplements products and active ingredients is not surprising when one considers the nature of many of these products. The DSHEA's expanded definition of dietary supplement to include herbs and botanicals, as well as concentrates, metabolites, constituents, and extracts of these plant products, resulted in a potentially unlimited number and variety of new chemical compounds meeting the legal definition of a "dietary supplement." Unlike vitamins and minerals, herbs and botanicals are not simple discrete chemical compounds. They are mixtures of perhaps hundreds of discrete chemical compounds. And there is absolutely no doubt that many of these plant-derived chemicals have potent medicinal as well as toxic potential. It is virtually impossible for nutritional scientists, toxicologists, and FDA regulators to adequately study and monitor or evaluate the safety and efficacy of all of these plant-derived dietary supplement ingredients. Moreover, recall that the DSHEA does not require that dietary supplement manufacturers conduct safety and toxicity studies on their products before marketing them. Even if the dietary supplement contains an NDI, the manufacturer is only required to provide the FDA with some evidence that the NDI is safe. But as we discussed earlier, that "evidence" standard is very weak; certainly much weaker than the FDA would expect for the safety evaluation of a new drug or a new food additive.

The challenges faced by health professionals and government health research and policy experts become apparent when you take a quick look at the "Dietary Supplement Fact Sheets" published by the Office of Dietary Supplements (ODS) within the National Institutes of Health (NIH).[15] Recall from an earlier chapter that the ODS was established as one of the provisions of the DSHEA. Its stated mission:

> *...to strengthen knowledge and understanding of dietary supplements by evaluating scientific information, stimulating and supporting research, disseminating research results, and educating the public to foster an enhanced quality of life and health for the U.S. population.*

[15] NIH, Office of Dietary Supplements. Dietary Supplement Fact Sheets. https://ods.od.nih.gov/factsheets/list-all/.

The ODS "Dietary Supplement Fact Sheets" website reviews available safety and effectiveness data on approximately 130 different dietary supplement substances, approximately 65 of which are botanicals, herbs, or substances derived from these sources. For the vast majority of these plant-derived products or ingredients, the ODS could only conclude that there was little evidence to support their effectiveness, and in several cases, there were serious concerns regarding safety.

The safety and effectiveness of botanical-derived dietary supplement products or ingredients is further complicated by the lack of validated and standardized methods for analyzing the chemical composition of these products. In addition, as is the case for any plant-derived substance, the quantity of the substance can vary widely depending on plant variety, part of the plant that is used (leaf, stem, root, etc.), growing conditions, processing conditions, and other factors. The result is a dietary supplement marketplace which too often contains products that do not contain the ingredients as stated on the label or that contain other fillers or adulterants. These issues drew national media attention by two fairly recent (2013) studies. In the first, Canadian researchers used DNA barcoding technology to test the composition of 44 herbal supplement products. They found that more than half of the products contained DNA from plants not listed on the label.[16] In a second study, also conducted in Canada, researchers reviewed 465 Class I drug recalls conducted in the United States between 2004 and 2012. They found that more than half of these recalls were for adulterated dietary supplement products that contained unapproved drug ingredients.

So while it may be reasonable for consumers to make informed decisions about vitamin and/or mineral supplements, it is virtually impossible for them to make similar informed decisions for other dietary supplement products for which there is insufficient scientific information related to safety and effectiveness, and which are of questionable composition and purity. Furthermore, in the absence of good scientific data on safety and effectiveness, how do manufacturers and consumers make informed decisions on dosing recommendations? For FDA-approved drugs (both prescription and OTC), manufacturers must provide the FDA with dosing information and recommendations that carefully consider the doses that are necessary to accomplish the intended benefit, against the doses that may result in toxicity. Of course, this can only be done following lengthy and expensive animal and human clinical testing. With this information, drug manufacturers employ a variety of scientifically sound

---

[16] Newmaster, SG, M Grguric, D Shanmughanandhan, S Ramalingam, S Ragupathy. DNA Barcoding Detects Contamination and Substitution in North American Herbal Products. *BMC Med.*, 11: 222 (2013).

strategies for making dosing recommendations. One such strategy is the use of the "Therapeutic Index" (TI) principle. The TI is simply a calculation that describes the ratio of the toxic dose of a substance in 50% of the population (TD_{50}) to the effective dose in 50% of the population (ED_{50}). That is:

$$TI = TD_{50}/ED_{50}$$

The larger this ratio, the "safer" the drug. In other words, a TI of "10" would mean that the dose that is toxic in 50% of the population is ten times higher than the dose that is effective in 50% of the population. Thus, a drug with a TI of 100 is "safer" than a drug with a TI of 10 (assuming all other factors are equal). When deciding on a recommended dose for a drug, manufacturers and regulators will typically consider the TI. There is no absolute acceptable value. It would depend on many factors, such as the intended purpose of the drug, the nature of the possible toxic effects, etc. For example, for a drug intended to treat a very serious, life-threatening disease, the FDA may be willing to consider a lower TI, because the higher risk of the drug is acceptable given the risk of not using the drug. On the other hand, for an OTC drug used to treat a non-serious health condition (perhaps a common cold remedy), the FDA would likely expect a relatively high TI value (relatively very low risk). The manufacturer or regulator could even consider a modified TI calculation that includes an additional margin of safety for the substance. For example, instead of looking at the TD_{50}/ED_{50} value, we could look at the TD_1/ED_{99} value. This value describes the ratio of the toxic dose in the most sensitive 1% of the population to the effective dose in the vast majority (99%) of the population.

But the use of even this most basic tool for establishing dosing recommendations is not possible with most dietary supplement products, due to the dearth of good scientific information on safety and effectiveness. How then can manufacturers make a label recommendation for the dose that a consumer should take of a dietary supplement? Well, for vitamin and mineral substances for which there are established RDA, AI, and UL values, these can be used as guidance for dosage recommendations. As noted earlier, however, these values have not been established for most other dietary supplement products. Manufacturers must therefore rely on what little information they do have regarding the safety and effectiveness of the product to make a "best guess" recommendation for dosage. For example, a green tea extract supplement dose may be based on the relative equivalent number of cups of actual green tea that the supplement supplies. But this is a rather "nonscientific" approach that can present its own unique risks. For example, when people eat or drink whole foods, such as fish, or fruits and vegetables, or even green tea, the amount that they consume is typically "self-limiting." That is, it would be rather difficult to consume (drink) so much actual green tea in a given day as to pose a toxic risk. However, when the components of the green tea are presented to the consumer in the form of an extract pill dietary supplement, it is much easier to consume too much, despite what the label recommends. This is particularly true when you consider that many consumers of these products incorrectly assume that "more is better."

So, clearly there are still many issues related to dietary supplement regulation that can be improved. As we consider how this may be accomplished in the future, let's summarize what we know, *in general*, about the safety and effectiveness of dietary supplements and related issues.

1. Due, at least in part, to the passage of the DSHEA in 1994, the number and variety of dietary supplement products on the market has increased enormously over the past two decades.
2. Based on assessments of available scientific data conducted by the NIH's ODS, the vast majority of dietary supplements on the market offer little to no health benefits. There are clearly a few that may offer some minor health benefit, but even for these, definitive studies are often lacking.
3. The vast majority of dietary supplement products on the market are relatively safe when used in moderation and according to label instructions. Again, however, the NIH's ODS identifies many dietary supplement products that present potentially significant health and safety concerns.
4. Some of the most serious health and safety concerns are typically associated with dietary supplement products for which the principle ingredients are "non-nutrient" in nature (that is, not a vitamin, mineral, protein, etc.). These would include supplements containing herbal or other botanical extracts, concentrates, or related synthetic-derived ingredients.
5. The scientific evidence related to the metabolism, biological functions, and safety of nutrient-based dietary supplements, as a class, is relatively strong. This allows health professionals and regulators to better advice consumers regarding RDAs, AIs, and ULs. Nevertheless, even for this class of dietary supplements, there are important examples of products that present significant health and safety concerns. For example, consider the concerns raised regarding the safety of amino acid supplements discussed in an earlier chapter. Or some of the recent studies that have identified increased risks of cancer or cardiovascular disease associated with the use of antioxidant supplements.[17]

---

[17] Goran, B, D Nikolova, L Lotte Gluud, RG Simonetti, C Gluud. Mortality in Randomized Trials of Antioxidant Supplements for Primary and Secondary Prevention: Systematic Review and Meta-analysis. *JAMA*, 297(8): 842–857 (2007).

6. The large number of bioactive and potentially toxic chemicals present in many plant-based supplements makes them extremely difficult to evaluate and regulate for safety and purity. As a result, many dietary supplements on the market do not contain the ingredients as stated on the label and/or contain adulterants not listed on the label.
7. Many dietary supplement manufacturers are including structure–function claims on the labeling and advertising of their products that are not backed by good scientific evidence and consensus.
8. Many consumers and health professionals are misinformed regarding the truthfulness of dietary supplement labeling claims and the role that the federal government (particularly the FDA) plays in monitoring and enforcing dietary supplement safety and health claims.

Addressing any of these issues and concerns through either legislative action or Executive Branch regulatory options would certainly be a challenge. The history of the regulation of dietary supplements discussed in the 15 chapters of this book is a testament to this challenge. And the passage of the DSHEA has left the FDA with severely limited regulatory options. As far as Congress passing any major changes to the DSHEA, it is worth noting that Congress passed the law in 1994 without even a recorded vote in either the Senate or House of Representatives. The political will to challenge the bill at that time was simply nonexistent, despite the rather weak science related to the effectiveness and safety of dietary supplements. It passed by voice vote in the Senate and without objection in the House. And the recent attempts by Senators McCain and Durbin to more rigorously regulate this industry never even made it out of committee. Still, passage of the Dietary Supplement and Nonprescription Drug Consumer Protection Act in 2006, and the FSMA in 2011 resulted in significant improvements to the regulation of this industry and demonstrated that legislative solutions to improve dietary supplement regulation are not impossible to achieve.

Of the eight issues and concerns listed earlier, perhaps the most difficult to address via the legislative approach may be the issue of how to deal with the safety of the myriad of potential active ingredients in herbal and other botanical supplements. Dr. Donald Marcus, professor emeritus at the Baylor College of Medicine (Houston, Texas) and an outspoken critic of the DSHEA, has suggested that the law needs to be revised or repealed with the goal of regulating herbal and other medicinal supplements as "medicines."[18] In other words, as drugs. But even he acknowledges that this would be very difficult to achieve. Short of changing the law, the FDA is working toward what is likely a more realistic and practical approach.

[18] Marcus, DM. Dietary Supplements: What's in a Name? What's in a Bottle. *Drug Test. Anal.*, 8: 410–412 (2015).

That is, to use its "NDI" regulatory authority (as discussed in Chapter ten). The agency's most recent draft of its "Guidance for Industry" on this topic[19] provides a significant amount of detail on what qualifies as an NDI and the evidence of safety that the agency would require of NDI's used in dietary supplements. While this approach could potentially identify and eliminate NDIs that pose significant health risks, it will certainly be a challenge for the FDA to enforce the NDI rules for the large number of products and product ingredients being introduced on the market each year.

Clearly, more also needs to be done regarding the issue of dietary supplement product/ingredient identity and purity. To the extent possible, the FDA needs to more rigorously enforce the dietary supplement CGMP regulations. The use of third-party verification/certification programs can also go a long way to maintaining consumer confidence in the products they purchase. With continued pressure on our Congressional representatives, perhaps this approach can eventually become part of the law. In the meantime, the industry itself should encourage its members to take advantage of these types of verification and certification programs, such as those offered by the United States Pharmacopeia.

And what can be done about misleading or even false health-related structure–function claims on some dietary supplements? Fortunately, the DSHEA did not weaken the FDA's authority regarding the use of explicit (authorized) health claims on food labeling, and the application of the "significant scientific agreement" standard for these types of health claims. But that has not prevented some unscrupulous dietary supplement manufacturers from abusing the structure–function labeling provisions of the DSHEA. The DSHEA does specify that these structure–function claims must "be truthful and not misleading" (21 USC § 343(r)(6)). But structure–function claims are not held to the same "significant scientific agreement" standard required of other food health claims. What, therefore, are the enforcement criteria for holding structure–function claims to the "truthful and not misleading" requirements of the law? In December of 2008, the FDA published its "Guidance for Industry: Substantiation for Dietary Supplement Claims."[20] According to this document, the FDA applies the same standard as used by the Federal Trade Commission (FTC) for its regulation of claims in dietary supplement advertising. That is, there needs to

---

[19] *Dietary Supplements: New Dietary Ingredient Notifications and Related Issues: Guidance for Industry (Draft Guidance).* U.S. Food and Drug Administration. August 2016. (www.fda.gov/Food/GuidanceRegulation/GuidanceDocumentsRegulatoryInformation/ucm257563.htm).

[20] *Guidance for Industry: Substantiation for Dietary Supplement Claims Under Section 403(r) of the Federal Food, Drug, and Cosmetic Act.* U.S. Food and Drug Administration. (www.fda.gov/food/guidanceregulation/guidancedocumentsregulatoryinformation/dietarysupplements/ucm073200.htm).

be *"competent and reliable scientific evidence"* in support of the claim. This is a much weaker standard than the *"significant scientific agreement"* standard expected of other food health claims. But it still requires good science to back the claim. The FDA's Guidance for Industry document goes on to describe what is typically expected to meet the *"competent and reliable scientific evidence"* standard.

> *...tests, analyses, research, studies, or other evidence based on the expertise of professionals in the relevant area, that has been conducted and evaluated in an objective manner by persons qualified to do so, using procedures generally accepted in the profession to yield accurate and reliable results.*

The FDA goes on to describe the types of studies that would be sufficient to substantiate a structure–function claim. Ideally, these include human randomized clinical trials (intervention studies) as providing the strongest evidence. But it is precisely these types of studies that the ODS of the NIH finds are often either lacking or fail to confirm a benefit for most of the dietary supplements on the market today. So while it may be laudable that the FDA and FTC have described scientifically sound substantiation standards for structure–function claims, the reality remains that there are many dietary supplements on the market today making structure–function claims that are based on much weaker scientific evidence.

It is also important to note that FDA "Guidance" documents (of which we have mentioned many throughout this book) do not have the same enforcement power as laws or regulations. Unlike FDA regulations, guidance documents have not gone through the formal rulemaking process required for new regulations and thus do not have the power of law. As the FDA notes on its website[21]:

> *Guidance documents represent FDA's current thinking on a topic. They do not create or confer any rights for or on any person and do not operate to bind FDA or the public. You can use an alternative approach if the approach satisfies the requirements of the applicable statutes and regulations.*

[21] *Guidances.* U.S. Food and Drug Administration. (www.fda.gov/forindustry/fdabasicsforindustry/ucm234622.htm).

In other words, these documents are, as their name implies, simply guidelines. Dietary supplement manufacturers are not legally obligated to follow the guidelines, as long as they still obey the applicable laws.

So where does that leave the consumer who is faced with trying to navigate the confusing array of dietary supplement label claims? Products that make exaggerated, misleading, or outright untruthful health claims have been around for hundreds of years. And no matter how confident we may feel we are at recognizing false or misleading claims, when faced with a challenging health issue or health goal, even the most skeptical consumer can fall prey to a slick label or advertisement. Still, skepticism is a very powerful defense against nutrition and health quackery. All good scientists are skeptics at heart. And all smart consumers should be, as well. The age-old maxim still holds true, especially in the world of nutrition and dietary supplements; *If it is too good to be true, it probably is.*

Knowledge and facts are also essential to being an informed consumer of health and nutrition products and services. However, it simply is not realistic nor reasonable to expect the average "nonscientist" consumer to be able to sort through, understand, and critically evaluate the enormous amount of nutrition information (and misinformation) that they are bombarded with on a daily basis. Yet without this ability, how can they be expected to be able to make informed purchasing decisions? Perhaps this is an area where the federal government can offer some assistance. Recall from our discussion of the structure–function claim provisions of the DSHEA in Chapter ten that the law requires that manufacturers notify the FDA of their intention to use a structure–function claim on their product labeling within 30 days of the product being released on the market. Recall also (from Chapter ten and our discussion earlier in this chapter) that the DSHEA requires that *"the manufacturer of the dietary supplement has substantiation that such statement is truthful and not misleading."* (21 USC § 343(r)(6)). So, even though the DSHEA does not stipulate that manufacturers provide this substantiation to the FDA as part of the 30-day notification, it does presume that the evidence to substantiate the claim exists from the manufacturer. Perhaps the dietary supplement labeling regulations could be amended to require that manufacturers whose products include a structure–function claim also include a web address (URL) on the label that links to a website describing the evidence that the company is using to substantiate the claim. Not only would this information help consumers make informed decision, but it would also allow nutritional scientists, other health professionals, and FDA regulators to critically evaluate the evidence and offer consumers a balanced and unbiased evaluation. The label could perhaps also include a link to the ODS (within the NIH) for more general consumer information and advice regarding the safety and efficacy of various dietary supplements.

In the end, the goal of all parties involved in the dietary supplement arena should be to balance a respect for consumers' right to free choice and access to health products, while ensuring that the products are safe, at least somewhat (or potentially somewhat) effective, and that are not labeled or advertised based on false or misleading information. What this book has hopefully demonstrated is that this is an extraordinarily difficult goal to achieve. It has now been well over a century since the federal government first stepped into the debate. As you consider all that has happened since the passage of the Pure Food and Drug Act of 1906, it may be tempting to conclude that we have not progressed very far from the "snake oil salesman" days of the 19th century. In some respects, I would tend to agree. But overall, I would suggest that we are moving in the right direction. I am optimistic that we can achieve this balance in stakeholder goals and that consumers will continue to become more knowledgeable and discerning when it comes to dietary supplements and the role they can (and cannot) play in personal health.

Index

Printed in the United States
by Baker & Taylor Publisher Services